STUDIES IN CHEMICAL PHYSICS

General Editor

A. D. Buckingham, Professor of Chemistry, University of Cambridge

Series Foreword

The field of science known as 'Chemical Physics' has greatly expanded in recent years. It is an essential part of both physics and chemistry and now impinges on biology, crystallography, the science of materials and even on astronomy. The aim of this series is to present short, authoritative and readable books on different topics in chemical physics at a level that is appreciated by the non-specialist and yet is of prime interest to the expert—in fact, the type of book that we all welcome and enjoy.

I was grateful to be given the opportunity to help plan this series, and warmly thank the authors and publishers whose efforts have brought it into being.

A. D. Buckingham,
University Chemical Laboratory,
Cambridge, U.K.

Other titles in the series

Advanced Molecular Quantum Mechanics, R. E. Moss
Chemical Applications of Molecular Beam Scattering,
K. P. Lawley and M. A. D. Fluendy

Electronic Transitions and the High Pressure Chemistry and Physics of Solids

Electronic Transitions and the High Pressure Chemistry and Physics of Solids

H. G. DRICKAMER AND C. W. FRANK

School of Chemical Sciences and
Materials Research Laboratory,
University of Illinois, Urbana

CHAPMAN AND HALL · LONDON

First published 1973
by Chapman and Hall Ltd
11 New Fetter Lane, London EC4P 4EE

© *1973 H. G. Drickamer and C. W. Frank*

Filmset by Santype Ltd, Salisbury, Wilts.
Printed in Great Britain by
Fletcher & Son Ltd, Norwich, Norfolk

SBN 412 11650 2

Distributed in the U.S.A.
by Halsted Press, a Division
of John Wiley & Sons, Inc.,
New York

Contents

Acknowledgements

The authors wish to acknowledge with gratitude permission to use figures from the following sources: Journal of Chemical Physics, the Physical Review, and Physical Review Letters published by the American Institute of Physics; Proceedings of the National Academy of Sciences published by the Academy; Science published by the American Association for the Advancement of Science; the Journal of Solid State Chemistry published by Academic Press; Journal of Physics and Chemistry of Solids published by Pergamon Press; 'Solid State Physics, Volume 17,' 'The Physics of Solids Under Pressure,' and 'Organic Charge Transfer Complexes' published by Academic Press.

Preface

There is no paucity of books on high pressure. Beginning with P. W. Bridgman's *The Physics of High Pressure*, books of general interest include the two-volume *Physics and Chemistry of High Pressure*, edited by R. S. Bradley, and the series, *Advances in High Pressure Research*, as well as the report on the Lake George Conference in 1960. Solid state physics is well represented by *Solids Under Pressure*, edited by Paul and Warschauer, by *Physics of Solids at High Pressure*, edited by Tomizuka and Emrick, and by *Properties Physiques des Solides sous Pression*, edited by Bloch, as well as by chapters in Volumes 6, 13, 17, and 19 of *Solid State Physics*, edited by Seitz, Turnbull, and Ehrenreich. Chemistry in gases and liquids is covered in Weale's *Chemical Reactions at High Pressure*, and Hamann's *Physico-chemical Effects of Pressure*. In addition to the coverage of techniques and calibrations in the above volumes, *Modern Very High Pressure Techniques*, edited by Wentorf, *High Pressure Methods in Solid State Research*, by C. C. Bradley, *The Accurate Characterization of the High Pressure Environment*, edited by E. C. Lloyd, and a chapter in Volume 11 of *Solid State Physics* are devoted entirely to this facet of high pressure research. It is not our plan either to supersede or extend these approaches.

It is our purpose here to discuss the effect of high pressure on the electronic properties of solids. We sharpen our focus further to a particular set of phenomena, but a set of very general occurrence. Pressure tends to shift the energy of one set of orbitals with respect to another; under a wide variety of circumstances this shift can lead to a new ground state, or a greatly modified ground state for a system. We define this event as an electronic transition. Bridgman demonstrated that polymorphic

ix

transitions – rearrangements of atoms, ions, or molecules in a crystal lattice – are common phenomena. In the past twenty years it has been demonstrated that electronic transitions are also ubiquitous, and that they have chemical as well as physical consequences. The physical consequences have been widely and thoroughly studied by the solid state physics community. We discuss these physical consequences, but we strongly emphasize the chemical effects which have not been so thoroughly explored. We also wish to point out that studies of the pressure effects on electronic orbitals can be important to an understanding of atmospheric pressure physics, chemistry, and biology.

It is difficult to select the appropriate time for a monograph in a given field. We feel that in the area covered here there has been sufficient progress to permit a reasonably well-rounded picture to emerge. Our approach has been frankly exploratory and speculative; we paint with rather a broad brush. We hope to interest experimentalists in many of the aspects which are barely scratched or completely untouched. We particularly hope to interest theoreticians in providing a deeper and broader understanding of the experimental observations.

HGD would especially like to acknowledge the collaboration of the many students who have made possible the parts of this work done in this laboratory. The interaction and collaboration with C. P. Slichter has been a vital factor in developing the concepts presented here. We owe a considerable debt to very enlightening conversations with R. A. Marcus. The critical analysis of the manuscript by D. W. McCall has been invaluable; we appreciate also comments by J. Jonas and H. Frauenfelder. Without the continuing encouragement of F. M. Beringer and H. S. Gutowsky this monograph would not have been written.

It is also a pleasure to acknowledge the continuing interest and support of the United States Atomic Energy Commission for many aspects of this work.

Urbana, Illinois　　　　　　　　　　　　　　　　H. G. Drickamer
April 1972　　　　　　　　　　　　　　　　　　　C. W. Frank

CHAPTER ONE

Introduction

The basic effect of pressure is to increase overlap between adjacent electronic orbitals. There are a number of consequences of this increased overlap. In the first order, there is a delocalization of electrons, a broadening of the bands of allowed energy and a decrease in magnitude of the gaps of forbidden energy between bands. For an insulator or semiconductor there is a decrease in resistivity which may ultimately lead to metallic behavior. This is the classical picture observed in books like that of Seitz [1].

A second effect of pressure is the relative displacement of one type of orbital with respect to another. Since orbitals with different quantum numbers may differ in radial extent, or orbital angular momentum (orbital shape), or in diffuseness or compressibility, one might expect that this would be a rather common phenomenon; we shall encounter many examples of it. In terms of the band picture of solids, the relative shift may act to augment or to oppose the broadening effect. These relative displacements also affect many of the excitations of interest in chemistry. Three classes of examples are emphasized in this monograph. One of these is the shift in energy of the empty π^* orbitals *vis-à-vis* the occupied π orbitals of aromatic hydrocarbons and other systems of conjugated π orbitals. A second is the change in relative energy of d orbitals of one symmetry with respect to d orbitals of different symmetry in transition metal complexes. The third is the change of energy of an electron acceptor with respect to a donor – either in a molecular electron donor-acceptor complex, or in a transition metal ion complex.

1

[*Refs. on p. 5*]

The third effect of interest is a result of the relative shifts discussed above. Under a wide variety of circumstances, there may be an excited state not too high in energy with respect to the ground state. Then the relative displacement of one type of orbital with respect to another may be sufficient to establish a new ground state for the system, or greatly to modify the properties of the ground state by configuration interaction. This event we call an electronic transition. As discussed further below, one of the most active and important aspects of current high pressure research is the discovery and understanding of electronic transitions as broadly defined above. This has been primarily an activity of the high pressure physics community, but we shall demonstrate that these electronic transitions have chemical consequences which may well be broader than the purely physical effects. The understanding of the generality and significance of electronic transitions rests on two experimental developments: the extension of the available range of pressure to the order of several hundred kilobars, and the development of techniques which permit the measurement of a number of electronic properties over this range. The latter development is the more important.

High pressure research is very largely based on the pioneering work of P. W. Bridgman [2], who developed most of the important techniques and made a variety of measurements of the macroscopic properties of condensed systems. Some twenty-five years ago, A. W. Lawson demonstrated that Bridgman's 12 kbar cell, utilizing a liquid or gaseous pressure transmitting medium, could be used for a wide variety of studies of the electronic properties of solids. Since that time, practically all of the usual experiments of modern solid state physics have been performed in the 12 kbar range. Bridgman's 30 kbar cell has also been used for solid state investigations.

Bridgman also developed a piston-cylinder device for pressure-volume measurements in solids to 50 kbar, as well as a more complex device for similar measurements in the 100 kbar range. In addition, he invented the opposed anvil configuration known as 'Bridgman anvils' which is discussed in more detail in Chapter 5. In the form Bridgman used it, he was able to make electrical resistance measurements to about 70 kbar. Since that time the belt apparatus and the multiple anvil apparatuses have been developed. These permit electrical measurements and provide relatively large volumes for chemical syntheses at pressures to 100 kbar or above over a considerable range of temperature.

The supported taper device used for many of the measurements discussed here has been applied to optical absorption, electrical resistance,

[*Refs. on p. 5*]

X-ray diffraction, and Mössbauer resonance studies using very small sample volumes, at pressures of several hundred kilobars.

In addition to his development of techniques, Bridgman, of course, made a large number of experimental measurements of importance to solid state science. In fact, his measurements of the volume and resistance of cesium as a function of pressure contributed to the concept of an electronic transition. Probably the result of his high pressure experiments of most general interest was that the polymorphic transition – a first order phase transition involving rearrangement of the atoms – is a very common phenomenon.

During the last decade it has been demonstrated that electronic transitions, as defined above, are also ubiquitous phenomena. Most of the effort has been towards demonstrating changes in physical properties, particularly in volume and in electrical conductivity. We wish to emphasize here the effects of changing the nature of the ground state on the chemical properties of solids.

Fig. 1.1 has been prepared to clarify the picture as to the nature of the events discussed here. We divide the types of transitions observed in solids into four classes. Class I refers to events which involve rearrangement of the atoms, ions, or molecules with no electronic implications. Class II contains events whose main feature appears to be

Class I First order transition – electronic component negligible	Class II First order transition – significant electronic component	Class III electronic transition – significant volume or structure change	Class IV electronic transition – continuous
fcc → sc KCl, KBr, KI	bcc → hcp Iron (Ferro → paramagnetic)	s → d Cesium Rubidium	reduction of Fe(III) to Fe(II)
fcc → hcp Lead	diamond → white tin Si, Ge, InSb, GaAs etc. (semiconductor → metal)	Potassium? f → d Cerium other rare earths	spin changes in partially filled shells reactive ground states of aromatic hydrocarbons and
		Mott Transition doped V_2O_3?	electron donor-acceptor complexes
			rare earth salts

Fig. 1-1 Classification of high pressure transformations.

[*Refs. on p. 5*]

polymorphism, but where there may be drastic changes in electrical or magnetic properties. Class III contains essentially electronic transitions which occur discontinuously and are accompanied by a volume discontinuity. Class IV involves new ground states established over a range of pressures, co-operative phenomena, or events involving continuous change in the degree of configuration interaction. The equilibrium between states here is like the equilibrium between reactants and products in any chemical reaction.

The boundaries between the various classes are, of course, somewhat arbitrary. Any rearrangement of atoms or ions implies some change in electronic interaction because of change of local symmetry. Whether the electronic component of a transition is major or minor is, in some degree, a matter of viewpoint. Whether a transition occurs continuously or discontinuously may depend on the temperature at which it is studied. Nevertheless, the classification affords a basis for understanding what phenomena are discussed in this monograph.

In the first place, we restrict ourselves to events in Class III and Class IV. We discuss both 'physical' events – typically those of Class III – and 'chemical' events. However, since our purpose is, in large part, to introduce these areas of science to a chemical audience, the emphasis is on electronic transitions with chemical consequences. One must keep in mind, however, that electronic transitions have implications for geophysics and biology, as well as for physics and chemistry. In many cases, the transformations have strong co-operative aspects, but the electronic energy levels can always be expressed in terms of atomic orbitals or combinations thereof. We do not discuss such basically co-operative phenomena as ferromagnetism, antiferromagnetism and superconductivity. Thus we do not cover the work of such scientists as Bloch, Tomizuka, Samara, and Wittig.

In Chapter 2 we review some of the essential aspects of molecular orbital theory, ligand field theory, the theory of electron donor-acceptor complexes, and the band theory of solids. These are covered in outline only, to put people with adequate background but from different disciplines on a common ground of ideas and nomenclature. In Chapter 3 we present a brief analysis of the relationship between the energies associated with optical and thermal electron excitation processes. This is not covered in standard texts but is essential to an understanding of electronic transitions. Chapter 4 presents a phenomenological analysis analogous to molecular field theories of magnetism or regular solution theory. We establish the co-operative nature of certain of these

[*Refs. on p. 5*]

transitions, and the circumstances under which they will occur continuously or discontinuously, as well as the cause of hysteresis.

Chapter 5 is a brief description of the types of measurements used. The technological and mechanical aspects are de-emphasized. The thrust of the chapter is to establish the possibilities and limitations of experimental work on electronic behavior for pressures beyond the 100 kbar range. In Chapter 6 we review experimental data on the relative shift of one set of orbitals with respect to another. The range of phenomena covered includes: d-d transitions in transition metal ions, excitation of F centers in alkali halides, π-π* transitions in aromatic and related molecules, charge transfer in molecular and transition metal complexes, and excitations from the valence to the conduction band in insulators and semiconductors.

Chapter 7 discusses electronic transitions in alkali, rare earth, and alkaline earth metals, as well as insulator-metal transitions. These are the events usually denoted as electronic transitions in the physics literature. The major observations are outlined, but no attempt is made to cover the rather vast literature completely. Chapter 8 covers changes in spin state in compounds of iron, while Chapter 9 discusses the reduction of ferric iron by pressure. In Chapter 10 we discuss molecules, including biological prototypes, where there are apparently changes in both oxidation state and spin state. In Chapter 11 we discuss new classes of chemical reactions in aromatic hydrocarbons and their molecular complexes at high pressure, caused by the establishment of a new reactive ground state.

References

1. F. SEITZ, *Modern Theory of Solids*, McGraw-Hill, New York (1940).
2. P. W. BRIDGMAN, *Physics of High Pressure*, G. Bell and Sons, Ltd., London (1949).

CHAPTER TWO

Theories of Electronic Energy Levels in Molecules and Solids

2.1 Molecular orbital theory

The most comprehensive available treatment for the analysis of the electronic behavior of molecules is that of molecular orbital (MO) theory. The theory has been used to describe many aspects of molecular structure and such diverse molecular properties as optical absorption spectra, electronic dipole moments, and electron and nuclear magnetic resonance. Numerous texts exist on the treatment of molecular orbital theory at various levels of approximation. Streitweiser [1] considers Hückel's π electron theory in detail. Later works by Salem [2] and Murrell [3] develop the self-consistent theory for π electron systems. Finally, Pople and Beveridge [4] consider more recent approximate molecular orbital theories which may be applied to all valence electrons of a general three-dimensional molecule.

The basic concept is that the electronic orbitals extend over the entire molecule, allowing for extensive delocalization of the electrons occupying the orbitals. Localized bonding, as evidenced by significantly large amplitudes of the wave functions in certain regions of the molecule, is included as a special case. Since the orbitals are characteristic of the molecule as a whole, symmetry properties of the wave functions in the molecular point group are very important. Proper attention paid to symmetry relations in the selection of zero-order wave functions often allows the elimination of most of the off-diagonal elements of the Hamiltonian matrix. In addition, symmetry governs selection rules so that recourse to explicit integration is not required. Tinkham [5] gives an

6   [Refs. on p. 31]

excellent description of symmetry and group theory and its application to quantum mechanics so only brief comments will be presented here.

A symmetry operator is any co-ordinate transformation which leaves the Hamiltonian invariant or, in other words, which leaves the molecule unaltered. Symmetry operations include rotations around an axis through an angle $2\pi/n(C_n)$, reflections in a plane of symmetry perpendicular to the main symmetry axis (σ_h), reflections in a plane containing the main symmetry axis (σ_d or σ_v), reflection in a center of symmetry (i), improper rotations consisting of a given rotation C_n followed by $\sigma_h(S_n)$, and finally the identity operation (E). Operations of the same physical kind may be grouped in classes; the complete set of such symmetry operations for a given molecule constitutes the molecular point group. The most important point groups for the systems examined in this discussion are the octahedral O_h and square planar D_{4h}. Several D_3 systems have also been examined but the effects investigated are sensitive more to the local symmetry of the transition metal ion than to that of the entire molecule, usually allowing analysis in terms of O_h.

Each symmetry element of a group is associated with a square matrix, with the complete set of matrices chosen such that matrix multiplication parallels the combination of symmetry operations. A group representation then consists of a particular set of all such matrices.

The dimensionality of the representation is designated by A or B for one, E for two, or T for three dimensions. A and B refer to characters (traces of the representation matrix) of $+1$ and -1, respectively, under the principal rotation. A numerical subscript is used to distinguish between representations. If the sign of the representation does not change under the inversion operation, it is designated by g; if its does, by u.

Selection rules may be readily determined using group theory. In these cases molecular integrals of the form $\langle \Psi_A | \hat{Q} | \Psi_B \rangle$ must be evaluated where Ψ_A and Ψ_B form bases for the irreducible representations of the molecular point group and $\hat{Q}$ is the transition moment operator. Such an integral will be non-zero only if the direct product $\Gamma_A \cdot \Gamma_Q \cdot \Gamma_B$ is, or contains, the totally symmetric representation A_{1g}. This will occur if the representation of the direct product of any two of the functions is, or contains, the same representation as that given by the third function.

In principle, a quantitative determination of the molecular orbitals for a particular system may be obtained from an *ab initio* solution of the Schrödinger equation. However, sufficiently accurate results are obtainable only for the very simplest systems due to the prohibitively large amount of computer time required. For this reason, various

[*Refs. on p. 31*]

approximate methods have been developed. A useful feature which may be applied to these semi-empirical treatments is the formulation of the molecular orbital functions in terms of the corresponding atomic functions, the simplest such approximation being a linear combination of atomic orbitals (LCAO). Thus, for a given basis set of atomic functions, ϕ_j, a particular one electron orbital Ψ_i is represented as

$$\Psi_i = \sum_j c_{ij} \phi_j$$

where the c_{ij} are numerical coefficients, either real or complex, which must be determined. The various types of molecular orbitals formed by such a process are distinguished by the symmetry of the orbital overlap. If the overlap is symmetric with respect to rotation about the connecting axis of two atoms (s-s, s-p_z, p_z-p_z, p_z-d_{z^2}, etc.), the resulting molecular orbital is of σ character. If a nodal plane results along the connecting axis (p_x-p_x, p_y-p_y), a π molecular orbital is formed. Fig. 2.1 shows the bonding and antibonding σ and π orbitals formed from overlap

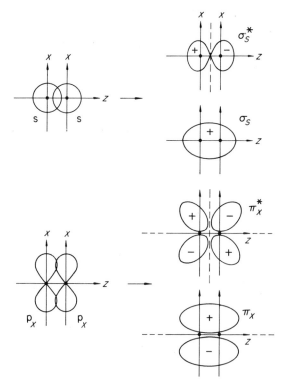

Fig. 2.1   Sigma and pi orbitals.

[*Refs. on p. 31*]

of two atomic s functions and two atomic p_x functions, respectively. The dashed lines indicate nodal planes. Thus σ orbitals have zero angular momentum around the chemical bond, and π orbitals unit angular momentum, so the analogy to atomic s and p orbitals is obvious.

For both saturated and unsaturated systems the σ electrons are essentially localized in bonds which establish the geometrical structure of the molecule. The π electrons, however, are delocalized over the entire molecule in conjugated systems due to extended p_π orbital overlap. Such delocalization is illustrated in Fig. 2.2(a) for a six-membered benzene ring. The geometrical structure is established by σ overlap of carbon sp^2 hybrid orbitals. The figure shows the lowest energy π molecule orbital formed from p_π atomic orbital overlap. Here the atomic orbital sizes are reduced for clarity of presentation. The only nodal plane is that of the plane of the molecule; the five energy molecular orbitals all have more nodal planes. Fig. 2.2(b) shows the ground state and first excited state of the aromatic molecule anthracene.

Because of the unique delocalization characteristics, a useful technique for understanding the chemical properties of aromatic systems is through the π electron approximation. In this method the π electron system is

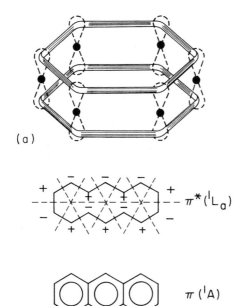

(a)

$\pi^*(^1L_a)$

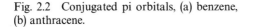

(b) $\pi(^1A)$

Fig. 2.2 Conjugated pi orbitals, (a) benzene, (b) anthracene.

[*Refs. on p. 31*]

treated explicitly with the σ system of electrons and the atomic nuclei considered to be part of a non-polarizable core. Of importance to high pressure experiments is the fact that the π overlap is usually considerably smaller than the σ overlap so that significant changes in the π electron structure would be expected upon application of pressure. The properties of the occupied and unoccupied π orbitals prove to be very important in the later discussion.

Of the semiempirical methods used for the determination of molecular orbitals, two different basic approaches are used, both of which originated in the framework of the π electron approximation. In the first, represented by the Hückel and extended Hückel methods, the elements of the inter-action energy matrix for the system are determined from essentially empirical considerations. The second approach, approximate self-consistent field theory, involves approximations to the atomic and molecular integrals in the energy matrix based on the mathematical formalism.

The simplest LCAO-MO approach is the well-known Hückel theory which, while crude, provides useful qualitative results. More sophisticated MO treatments for many electron systems are found in the self-consistent field Hartree-Fock and Roothaan methods; the Roothaan technique actually involves an approximation to the Hartree-Fock approach.

Although the Hartree-Fock approach yields the optimum molecular orbitals for small closed shell systems, solutions for larger molecules become impractical and further approximations must be used. The best approach to date has been the use of linear combinations of atomic orbitals to approximate the Hartree-Fock orbitals within the determi-nantal wave function. Virtually any desired accuracy is attainable by appropriate adjustment of the number of basis functions considered in the LCAO expansion. In the Roothaan approach the electronic energy contains elements of the charge density matrix arising from the products of the c_{ij}. The analysis of such charge density matrix elements and atomic overlaps constitutes a population analysis such as has been extensively developed by Mulliken [6].

The MO theory discussed thus far assumes that any state of a molecule can be described by a single many-electron configuration. Even if one were to use the best available MO basis set, such as the self-consistent field functions, the Slater determinant would only be a first approximation to the electronic wave function due to its neglect of electron correlation effects produced by Coulomb repulsion between

[*Refs. on p. 31*]

electrons of opposite spin. This neglect tends to increase the probability of finding two electrons in the same volume.

It is possible to account for the electron correlation effects by including interactions between states of different electron configurations but the same symmetry species. This corresponds to setting up many-determinantal functions, formed from different spin-orbitals but with the same total spin and orbital quantum numbers, and then taking linear combinations of these functions, with the coefficients chosen so as to minimize the energy. In the simplest case of interaction between two configurations, for example between the ground and lowest excited states of the same symmetry in a particular molecule, the two unperturbed levels will repel each other. The ground state will be lowered and the excited state raised in energy. The extent of configuration interaction is inversely proportional to the energy separation between unperturbed states. Thus the effect is particularly important for excited states since their energy levels are generally closer together. It should be noted that there are usually sufficient vibrational interactions in the complex solid state systems considered in this discussion to relax any symmetry restrictions related to configuration interaction.

Advanced methods of MO calculations employing configuration inter-action generally take one of two approaches. In the first case, rather arbitrary LCAO molecular orbitals, such as Hückel functions, may be used and then a considerable amount of configuration interaction is included to get good agreement with experiment. On the other hand, self-consistent field molecular orbitals may be used to set up the Slater determinants, in which case only a modest amount of configuration interaction is necessary.

2.2 Ligand field theory

The question to be answered in the study of co-ordination compounds is how to describe and characterize the bonding between the central transition metal ion and the surrounding ligands. Although the most comprehensive method of description is molecular orbital theory, one can get a clear qualitative picture from crystal field or ligand field theory, so we shall describe these first. The three theories may be traced to Bethe's classic paper [7] in which group theory was used to predict the electronic states resulting from the introduction of a given electronic configuration into an environment of less than spherical symmetry. The treatments differ in the extent and type of interaction between the central

[*Refs. on p. 31*]

metal and surrounding ligands. Whereas crystal field theory admits of no mixing of metal and ligand orbitals and assumes purely electrostatic interactions between point charges, ligand field theory allows a degree of covalent bonding such that the orbitals are no longer of pure metal or pure ligand character. However, though these theories give a good pictorial representation, semiquantitative calculations can be made, even for the fluorides, only by using molecular orbital theory with configuration interaction. Furthermore, molecular orbital theory is necessary for even a qualitative description of strongly covalent molecules, such as the cyanides or many organo-metallic compounds. Ligand field theory, including the application of MO theory, has been extensively discussed in books by Ballhausen [8] and Figgis [9] so that only the essential features will be outlined. This discussion will take the form of brief comments on the theory of atomic spectra followed by an outline of the crystal field through the molecular orbital treatments.

The energy levels for n equivalent electrons of a free ion in a central field may be determined by the techniques of Condon and Shortley [10]. The single-electron approximation is used with the hydrogen-like orbitals as a basis set. In this approximation the only total many-electron wave function which obeys the antisymmetry properties required is the Slater determinant. If the core-outer shell exchange is neglected, the closed argon shell of the first transition series may be ignored and attention focused on the $(3d)^n$ portion. For the case of no interaction among the d electrons, all d orbitals have the same energy. The degeneracy is lifted by successive considerations of electronic repulsion, spin-orbit interaction and the effect of external magnetic fields.

Most transition metal atoms exhibit Russell-Saunders or LS coupling for which the electron-electron repulsion energy is large compared to the spin-orbit interaction. If only repulsive effects are considered, an atomic configuration may be characterized by the two good quantum numbers L and S which specify the total orbital and total spin momentum obtained from coupling of all the orbital momenta and all the spin momenta separately. The projections of $\mathbf{L}$ and $\mathbf{S}$ on the z axis, m_L and m_S, are also good quantum numbers. Rather lengthy procedures, such as given in Condon and Shortley, exist for determining the terms arising from electronic configurations of equivalent d electrons. Such procedures ensure that the exclusion principle is not violated in specifying the allowed states. The term energies in this first approximation may be expressed in terms of the Condon-Shortley parameters, F_k, which are two-electron radial-type integrals and are positive by definition. This level of approximation

[*Refs. on p. 31*]

allows a confirmation of Hund's rule, i.e., for the terms arising from equivalent electrons, those with greatest multiplicity $(2S + 1)$ lie deepest, and the lowest is that with the greatest L. Discrepancies in the predictions of the ordering and energy of the excited terms arise from breakdown of the one-electron approximation and admixture of higher terms such as $(3d)^{n-1}(4s)$ into the configuration. Racah [11, 12] has defined new parameters $B = F_2 - 5F_4$ and $C = 35F_4$ which exhibit the special feature that the energy separation between terms with the maximum value of S is a multiple of B. These interelectronic repulsion parameters yield information about the change of radial extent of the d orbitals in pressure experiments.

The second level of approximation is to consider the spin-orbit interaction between the magnetic dipoles set up by the electron spin and orbital movement. The spin-orbit operator is treated as a perturbation on the electron repulsion scheme. Here m_L and m_S are no longer good quantum numbers, but their sum $m_J = m_L + m_S$ is. Thus the total angular momentum $\mathbf{J} = \mathbf{L} + \mathbf{S}$ is a good quantum number. This spin-orbit interaction splits the LS terms into $2S + 1$ levels characterized by J. The standard nomenclature for the term symbol of a given state is $^{(2S+1)}L_J$ where $L = 0, 1, 2, 3, 4 \cdots$ is designated by S, P, D, F, G, $\cdots$. For a shell less than half full, the state with the highest value of J has the highest energy; the reverse is true for a shell more than half full. A $2J + 1$-fold degeneracy still exists after consideration of spin-orbit interaction since the term energies are independent of m_J. This degeneracy is removed by a magnetic field, as exhibited in the Zeeman effect.

In crystal field theory the behavior of the electrons of the central metal ion is due solely to the electrostatic potential field of the surrounding ligands which possesses the appropriate molecular symmetry. Such a crystalline field lifts the degeneracy existing for the free ion in spherical symmetry. The role of the ligands is quite limited in this treatment, as the usual method of representing the field of six point charges by an expansion in spherical harmonics assumes that the potential obeys Laplace's equation and thus that the d orbitals do not overlap the charges. It is the breakdown of this assumption that precludes quantitative calculations based on the crystal field treatment.

The theory involves a Hamiltonian which differs from the free ion Hamiltonian considered earlier only in the inclusion of an electrostatic potential perturbation term. Three regimes may be distinguished through comparison of the size of the electrostatic crystal field potential with the other two perturbing quantities considered earlier, the electrostatic

[*Refs. on p. 31*]

repulsion and spin-orbit terms. First, rare earth complexes show larger spin-orbit than crystal field contributions. Second, the crystal field term for transition metal complexes in the weak field case is intermediate between the higher repulsive and lower spin-orbit terms. Finally, the strong field case exhibits a larger crystal field than repulsive contribution.

The presence of an octahedral crystal field has no effect on the s and p one-electron levels which transform as A_{1g} and T_{1u}, respectively. However, the five-fold orbital degeneracy of the d level is partially removed, as the five d orbitals span a reducible representation in the O_h group containing the irreducible representations E_g and T_{2g}. The three d orbitals transforming as T_{2g} are the d_{xz}, d_{yz}, and d_{xy} functions; the two transforming as E_g are d_{z^2} and $d_{x^2-y^2}$. Generally, the upper case symbols E_g and T_{2g} will be used to signify group representations and the states resulting from term splitting of the free ion; lower case symbols e_g and t_{2g}, on the other hand, represent single electron states. Group theory is of no aid in determining the ordering of the e_g and t_{2g} levels but consideration of repulsive forces makes it clear that the t_{2g} orbitals pointing between the negatively charged ligands will be more stable than the e_g set pointing directly at the ligands on the corners of the octahedron. By convention, the crystal field splitting is denoted by $10Dq$ or Δ as seen in Fig. 2.3. Simple electrostatic theory predicts an R^{-5} dependence for Δ. Generally, Δ is treated as a semiempirical parameter. However, an approximate relationship between the crystal field splittings for different cubic symmetries may be obtained from the point charge model. This gives $-9/4\,\Delta\,(\text{tetrahedral}) = \Delta\,(\text{octahedral}) = -9/8\,\Delta\,(\text{cubal})$.

If the symmetry is reduced there are more splittings and more free parameters. A common lower symmetry exhibited by several systems considered in this discussion is D_{4h}. This arises from a square planar arrangement or distortion of an octahedral structure by elongation or compression along one of the four-fold axes. The representation arising from the s level in D_{4h} symmetry remains A_{1g} but the p level is split into A_{2u} and E_u while the d level spans A_{1g}, B_{1g}, B_{2g} and E_g representations. The p orbital transforming as A_{2u} is p_z while p_x and p_y transform as E_u. The correspondence between the d orbitals and group representations is $A_{1g}(d_{z^2})$, $B_{1g}(d_{x^2-y^2})$, $B_{2g}(d_{xy})$, and $E_g(d_{xz}, d_{yz})$. Instead of one parameter to describe the splitting between the d orbitals, as in the octahedral case, three are required in square planar symmetry. If the departures from octahedral symmetry are small, the a_{1g} and b_{1g} levels resulting from the splitting of e_g in O_h will lie above the b_{2g} and e_g levels arising from the t_{2g} level in O_h. A reasonable scheme for the relative

[*Refs. on p. 31*]

(a) O_h

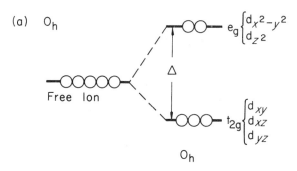

(b) D_{4h}

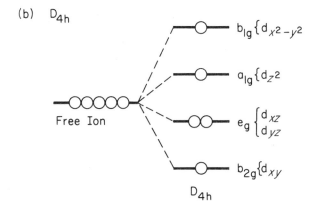

Fig. 2.3   Splittings in O_h and D_{4h} symmetry.

ordering of a_{1g} versus b_{1g} and b_{2g} versus e_g based on consideration of repulsive forces is given in Fig. 2.3. However, it should be noted that the relative ordering of the e_g and b_{2g} levels for D_{4h} symmetry may vary depending on the particular system.

Diagrams such as Fig. 2.3 are useful for visualizing orbital occupation. However, the energy states resulting from various d orbital configurations are best illustrated by a correlation diagram which relates the weak and strong crystal field cases. Quantitative correlation diagrams for d^n systems in the presence of both interelectronic repulsions and crystal fields of medium strength have been prepared by Tanabe and Sugano [13, 14]. These diagrams, in which the energy and crystal field splitting are given in units of B, are useful for evaluating Δ, B, and C from experimental spectra if the assignments are known. Although the diagrams are prepared for a specific C/B ratio, it is a straight-forward process to generate a new set for a different ratio by solving the secular determinants for the energy matrices on a high speed computer.

[*Refs. on p. 31*]

The metal ion in a transition metal complex may exhibit two or more spin states depending upon the strength of the ligand field and the molecular symmetry. Griffith [15] has shown that only two spin states are possible for octahedral crystal fields; however, fields of reduced symmetry, such as is exhibited by the D_{4h} square planar systems, lead to an intermediate state as well as high and low spin states. The orbital occupation scheme for the possible spin configurations of d^6 in O_h is given in Fig. 2.4. Other spin configurations are discussed in Chapter 10. The existence of a given system in a particular spin state depends on the relative values of the crystal field splitting and the mean spin pairing energy, Π. The spin pairing energy is determined from the increase in Coulomb repulsion energy and the loss of exchange, inherent in the pairing process. Griffith [15, 16] has calculated the mean spin pairing energy for d^6 (Fe^{2+}) to be $\Pi = 5/2\ B + 4C$. Thus a high spin state ($S = 2$) will result for $\Delta < \Pi$ and a low spin state ($S = 0$) for $\Delta > \Pi$. However, the crossover region is rather broad, covering a range of about 2000 cm^{-1}, due to the neglect of spin-orbit coupling and configuration interaction in the calculation. The possibility of a spin transition with increasing crystal fields is demonstrated clearly by the schematic correlation and semiquantitative Tanabe-Sugano diagrams. For the d^1, d^2, d^3, d^8, and d^9 configurations the ground state in the weak field limit is the

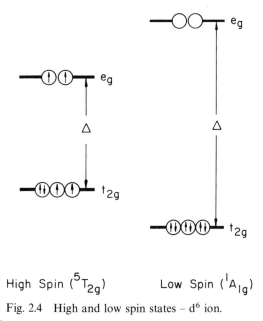

High Spin ($^5T_{2g}$)   Low Spin ($^1A_{1g}$)

Fig. 2.4   High and low spin states – d^6 ion.

[*Refs. on p. 31*]

same as in the strong field limit, implying no relative energy gain upon reaching the strong field configuration as the ligand field splitting increases. However, the ground state for d^4, d^5, d^6 and d^7 are not the same in the weak and strong field limits, the multiplicity being reduced as the crystal field potential energy overbalances the spin pairing energy. For example, the octahedral weak and strong field ground states for d^5 (Fe^{3+}) and d^6 (Fe^{2+}) are $^6A_{1g} - {}^2T_{2g}$ and $^5T_{2g} - {}^1A_{1g}$, respectively. The change of multiplicity indicates the formation of two additional sets of paired spins in each case, as is illustrated for d^6 in Fig. 2.4. The change of ground state is indicated on the Tanabe-Sugano diagrams by a discontinuity since the zero of energy is always taken as that of the lowest term. An estimate of the ease of spin pairing may be made from the location of the point of discontinuity. For d^5 systems this occurs at $Dq/B \cong 2.8$ and for d^6 at $Dq/B \cong 2.0$ for a C/B ratio of about 4.5. Depending upon the system, it is possible for pressure to effect a high to low spin or a low to high spin transition. The experimental evidence for these phenomena as well as the causes are discussed in Chapters 8 and 10, although the increase in crystal field splitting with pressure makes the possibility of a high to low spin transition readily apparent.

The discussion to this point has emphasized the use of symmetry to predict energy level splittings. In such a treatment the nature of the ligand is immaterial; only the symmetry of the electrostatic field is important. However, in order to describe experimental spectra, covalency properties which vary from ligand to ligand must be considered. Numerous experimental observations have been combined into two series which reflect the influence of various ligands on ligand field parameters. The first of these is the spectrochemical series in which typical ligand groups are ranked in order of increasing ligand field splitting for a constant metal ion: $I < Br < NCS^* < F < urea < OH < CH_3COO < oxalate < H_2O < SCN^* < glycine \sim ethylenediamine < tetracetic\ acid < pyridine \sim NH_3 < ethylenediamine < SO_3 < dipyridyl < O\text{-phenanthroline} < NO_2 < CN$. (The asterisk denotes the atom bonded to the metal). A second such series may be prepared on the basis of differences in the interelectronic repulsion parameter B. It has been observed that, for many systems, the interelectronic repulsion parameters for metal ions in complexes are less than for the corresponding gaseous ions. Thus B (complex) $= \beta B$ (free ion) where β is the 'nephelauxetic ratio.' This decrease in B, the so-called 'nephelauxetic effect,' has been reviewed by Jorgensen [17]. A portion of the nephelauxetic series is, in order of increasing value of B, diethyldithiophosphate $< I < Br\ CN \sim Cl < SCN^*$

[*Refs. on p. 31*]

< oxalate ~ ethylenediamine < NH_3 < urea < H_2O < F. Both series may be crudely rationalized on the basis of electronegativities, polarizabilities, polarizing power, permanent dipole moment, and π bonding characteristics but the details will not be pursued here. Physically, the decrease in interelectronic repulsion represented by decreasing β involves two factors: the spreading of the 3d orbitals which increases the average interelectronic distance and shielding of one 3d electron from another by ligand electrons.

It is of interest to consider covalency in general and mention two different types, central field and symmetry restricted covalency. The essential feature which the covalency theories attempt to explain is the expansion or delocalization of the d orbitals, causing an increase in the electron-electron distance and a decrease in the interelectronic repulsion. An electrostati⌐ mechanism for 3d expansion involves overlap of the metal electron cloud with the negative charge of the surrounding ligands. This reduces the effective nuclear charge felt by the outer portions of the d functions, causing them to expand. Another mechanism for 3d expansion involves 4s augmentation on the metal. The 4s electron density partially shields the 3d orbitals, causing them to expand. These two effects, which influence metal σ and π orbitals to the same extent, constitute central field covalency. Symmetry restricted covalency, on the other hand, acts differently on orbitals with different symmetries. This may involve metal d_π backbonding to vacant ligand π^* orbitals, which will be discussed later, or overlap of ligand electron density in the bond region. Due to the importance of these covalency effects, a more realistic approach to the analysis of the behavior of transition metal complexes is to consider molecular orbitals for the system composed of the central ion and the surrounding ligands.

Usually the molecular orbitals are formulated as linear combinations of atomic orbitals (LCAO) as indicated earlier in this chapter. Thus the wave function for a co-ordination compound is written as $\Psi = a_M \psi_M + \Sigma b_L \psi_L$ where ψ_M and ψ_L are atomic functions of the central metal ion and the ligands, respectively. A complete mathematical treatment can become immensely complicated due to the many-center integrals involved. However, use of an effective one-electron Hamiltonian has proven very helpful in the analysis. In the molecular framework the basis set of interacting orbitals will be limited to the outer shell atomic functions of the central metal ion and the ligands. In contrast to crystal field theory where only the five nd functions were considered, the MO treatment includes the $(n + 1)$s and the three $(n + 1)$p functions. The ligand contri-

[*Refs. on p. 31*]

bution consists of σ and π orbitals formed from appropriate combinations of ligand s and p functions. From these atomic functions suitable linear combinations which represent σ and π bonding and antibonding and π non-bonding orbitals for the molecule as a whole may be formed. The resulting schematic MO combination diagram for octahedral symmetry is shown in Fig. 2.5. Note that the highest partially filled orbitals, t_{2g} and e_g, are antibonding with π and σ character, respectively. They correspond in crystal field theory to the separated t_{2g} and e_g levels arising from the central ion d orbitals, although they now also contain some ligand character. Below the e_g and t_{2g} antibonding levels are the filled t_{1g}^n and t_{2u}^n levels which are π non-bonding and totally of ligand origin. The remaining filled levels, $t_{1u}(\pi)$, $t_{2g}(\pi)$, $e_g(\sigma)$, $t_{1u}(\sigma)$, and $a_{1g}(\sigma)$ have a large-to-moderate ligand contribution [9].

It is of interest to consider how the substitution of a slightly different ligand, or the change in ligand properties due, for example, to increase of pressure, affects the metal t_{2g} and e_g levels. First, an

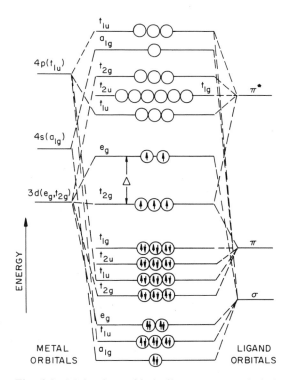

Fig. 2.5 Molecular orbital diagram – octahedral complex.

[*Refs. on p. 31*]

increase in σ donor ability of the ligand will increase the electron density in the bond region, causing the e_g level to increase in energy and the corresponding ligand σ level to decrease. As a result of the shift of the e_g level, the ligand field splitting will increase. On the other hand, a decrease in σ donor ability has the opposite results. The second point concerns the π bonding behavior. In many systems, particularly the cyanides, there are vacant π^* orbitals to which it is possible to delocalize some of the metal $t_{2g}(\pi)$ electrons. This π backbonding will decrease the t_{2g} orbital energy, increase Δ, and increase the ligand π^* level. This effect has important consequences which will be considered in the later discussion of actual systems.

Transitions between energy levels such as those already described for transition metal complexes may be made by either radiative or non-radiative (thermal) mechanisms. Only radiative processes will be considered here. Optical and thermal processes will be compared later in Chapter 3. First, the transition moment integral is examined to obtain the selection rules for the electric dipole transition mechanism, by far the most predominant of the multipole mechanisms. The probability of the transition is proportional to the transition moment integral $Q = \langle \Psi_1 | \mathbf{r} | \Psi_2 \rangle$ where Ψ_1 and Ψ_2 represent the total eigenfunctions of the two states and $\mathbf{r}$ is the radius vector having the symmetry of the electric dipole. In the Born-Oppenheimer approximation the total eigenfunction Ψ may be expressed as the product of electronic (ψ_e), vibrational (ψ_v), rotational (ψ_r), translational (ψ_t), and spin (ψ_s) wave functions. The rotational and translational contributions remain constant over the lifetime of an excited electronic state so they need not be considered. Thus, in the absence of spin-orbit interactions and vibronic coupling, the electronic, spin, and vibrational functions may be separated to yield the transition moment integral

$$Q = \langle \psi_{e1} | \mathbf{r} | \psi_{e2} \rangle \langle \psi_{s1} | \psi_{s2} \rangle \langle \psi_{v1} | \psi_{v2} \rangle. \tag{2.1}$$

The two major selection rules arise from the behavior of the first two integrals in Q. The first integral will vanish unless the direct product of the irreducible representations of ψ_{e1}, $\mathbf{r}$, and ψ_{e2} contains the totally symmetric representation. In the special case of centro-symmetric molecules it is clear that the two electronic states must be of unequal parity (gerade-ungerade character) since the radius vector is ungerade. Thus $g \rightarrow g$ and $u \rightarrow u$ transitions are parity or Laporte forbidden. Generally this is not a severe restriction for there is always some coupling of electronic and vibrational wave functions. This vibronic interaction

[*Refs. on p. 31*]

causes symmetry- or parity-forbidden transitions to have moderate intensity. The orbital selection rule may also be relaxed by 'intensity stealing' which is possible if the forbidden excited term lies near a fully allowed transition. Generally, a vibrational mode will exist such that the electronic wave functions of the excited term and the allowed level will mix, causing an increase in the intensity of the forbidden transition. This effect decreases as the energy difference between the excited term and the allowed level increases. Finally, the orbital selection rule may also be relaxed as a result of decreases of symmetry from the cubic case.

For the second selection rule it is noted that the two spin functions are orthogonal unless the two states they represent have the same multiplicity. Hence the spin selection rule is $\Delta S = 0$. This is much more rigorous than the symmetry selection rule, resulting in very low intensity for spin-forbidden transitions. Strong spin-orbit coupling will relax this rule, however.

The intensity of electronic transitions is generally described in terms of the oscillator strength which classically represents the number of electrons set into oscillation under the action of the radiation field. It is proportional to the frequency of absorption times the square of the transition moment and may be experimentally measured as the area under the curve for a plot of extinction coefficient ε versus the frequency v. Completely allowed bands have a maximum extinction coefficient of about 10^5 while spin-forbidden transitions have ε_{max} about 10^{-1}.

With this background on electronic spectroscopy, specific examples may be examined in more detail. The remainder of this section will be devoted to distinguishing between the types of optical absorption spectra. Optical absorption spectra may be classified into three groups. First, transitions between molecular orbitals localized on the central metal ion, $t_{2g}(\pi)$ and $e_g(\sigma)$, lead to d-d or ligand field bands. Second, transitions between molecular orbitals mainly localized on the ligands and molecular orbitals mainly localized on the central metal lead to ligand-to-metal or metal-to-ligand charge transfer bands. Finally, intra-ligand transitions may occur which are essentially unaffected by co-ordination to the metal.

The ligand field spectra will be considered using the d^5 configuration of Fe(III) as an example. For high spin d^5 octahedral systems the lowest lying ligand field transitions are predicted from the correlation diagram to be $^6A_{1g} \rightarrow {}^4T_{1g}$ and $^6A_{1g} \rightarrow {}^4T_{2g}$. Since the parity and spin selection rules are violated, these are of extremely low intensity.

In contrast to the ligand field bands which may be rather easily assigned using ligand field theory, the charge transfer bands are rather

[*Refs. on p. 31*]

difficult to interpret due to the difficulties in making the required molecular orbital calculations. Ligand-to-metal charge transfer bands (LMCT) occur in the visible or ultra-violet region and are exhibited particularly by complexes containing highly reducing ligands. These bands tend to shift to lower energy as the central metal becomes more oxidizing or the ligands more reducing. For equivalent ligands, the donated electron is considered to come from a delocalized ligand molecular orbital. Thus the predominant configuration for the excited state consists of the reduced metal and a collectively oxidized set of ligands. Four types of LMCT bands may be distinguished using molecular orbital configurations. They are $\pi_L \to \pi_M^*(t_{2g})$, $\pi_L \to \sigma_M^*(e_g)$, $\sigma_L \to \pi_M^*(t_{2g})$, and $\sigma_L \to \sigma_M^*(e_g)$ where the subscripts L and M refer to predominantly ligand and metal orbitals, respectively. For octahedral complexes, the σ_L molecular orbitals arise from the a_{1g}, e_g, and t_{1u} ligand orbitals while the π_L group includes the t_{1g}, t_{2g}, t_{2u} orbitals.

The reverse direction of charge transfer, metal-to-ligand (MLCT), is most likely to occur in complexes with central metal ions having small ionization potentials and ligands with easily available π^* orbitals. The transitions here may be represented as $\pi_M^*(t_{2g}) \to \pi_L^*$ and $\sigma_M^*(e_g) \to \pi_L^*$. The CN^- and CO ligands are examples where backbonding to empty π^* orbitals may occur. In addition, MLCT transitions are favored for a low oxidation state of the central metal ion. Note that for some systems both ligand-to-metal and metal-to-ligand charge transfer are possible.

Finally, intraligand bands resulting from transitions between two molecular orbitals which are both principally localized in the ligand system are considered. The types of transitions are $\sigma_L \to \sigma_L^*$, $n_L \to \sigma_L^*$, $n_L \to \pi_L^*$, and $\pi_L \to \pi_L^*$. Generally the $\sigma_L \to \sigma_L^*$ transition energy is very high, in the vacuum ultraviolet region. Molecules that contain non-bonding electrons on oxygen, nitrogen, sulfur, or halogen atoms exhibit $n_L \to \sigma_L^*$ transitions in the ordinary ultra-violet region. Still lower energy transitions are found for transitions to π_L^* orbitals when there are unsaturated centers in the molecule. The $\pi_L \to \pi_L^*$ transition is frequently of higher energy than the $n_L \to \pi_L^*$ transition.

2.3 Electron donor-acceptor complexes

Forces of interaction in molecules may range from the strong interactions which result in covalent or ionic bonds to the very weak interactions which may be explained in terms of dispersion, dipole-dipole, dipole-induced dipole and similar van der Waals forces. Intermediate between these extremes are the molecular complexes which result from generally weak interaction between molecules or ions which have electron donor

[*Refs. on p. 31*]

or acceptor character. Such complexes exhibit a definite stoichiometry, although a 1 : 1 overall ratio in the solid state does not necessarily imply pairwise units. These complexes have often been described as 'charge transfer' complexes although the major contribution to the binding forces in the ground state does not arise from transfer of charge. However, an intermolecular charge transfer electronic absorption often occurs which involves electron transfer from the donor to the acceptor. Much work has been done in the field of molecular donor-acceptor complexes but the main stimulus for the developments in the last two decades has been the work of Mulliken [18-20]. Recent books by Mulliken and Person [21] and Foster [22] review the theory and experimental observations.

Electron donor-acceptor systems have been classified according to the strength of the DA bond and on the effect of the bond on the internal chemistry of the donor or acceptor (e.g. good $b\pi$ donors, good $a\pi$ acceptors, etc.). Here we discuss results primarily for complexes of aromatic hydrocarbons (good $b\pi$ donors) with chloranil or tetracyanoethylene (π acceptors) or with I_2 (an $a\sigma$ acceptor).

We shall also be concerned with the self-complexing properties of aromatic hydrocarbons. This appears to be an important feature of the larger polycyclic aromatics.

The theoretical description of molecular complexes may be treated in terms of a valence bond, molecular orbital, or free electron approach. The comments in this discussion will concentrate on the valence bond and free electron treatments. The valence bond resonance approach due to Mulliken [18-20] will be considered first. Weak electron donor-acceptor inter ctions between molecules in their totally symmetric ground states may be described in terms of a wave function of the form

$$\Psi_N(AD) = a\,\Psi_0(A, D) + b\,\Psi_1(A^- - D^+) \qquad (2.2)$$

where $\Psi_N(AD)$ represents the complex, $\Psi_0(A, D)$ is the 'no bond' function corresponding to purely physical interactions, and $\Psi_1(A^- - D^+)$ is the dative function representing complete transfer of an electron from the donor to the acceptor. Generally b/a is very small, i.e., the charge transfer structure makes only a very small contribution to the ground state.

From second order perturbation theory the ground state energy W_N is found to be

$$W_N = \langle \Psi_N | H | \Psi_N \rangle = W_0 - X_0 \qquad (2.3)$$

$$= W_0 - \frac{(H_{01} - S_{01}\,W_0)^2}{(W_1 - W_0)} \qquad (2.4)$$

[*Refs. on p. 31*]

where H is the exact Hamiltonian, X_0 is the stabilizing resonance energy,

$$W_0 = \langle \Psi_0|H|\Psi_0\rangle, \quad W_1 = \langle \Psi_1|H|\Psi_1\rangle,$$

$$H_{01} = \langle \Psi_0|H|\Psi_1\rangle \quad \text{and} \quad S_{01} = \langle \Psi_0|\Psi_1\rangle.$$

The excited state $\Psi_E(AD)$ may be represented by

$$\Psi_E(AD) = a^{\ddagger}\Psi_1(A^- - D^+) - b^{\ddagger}\Psi_0(A, D) \tag{2.5}$$

where $a^{\ddagger} \cong a$ and $b^{\ddagger} \cong b$; these would be true equalities if S_{01} were zero.

The excited state energy W_E is given by

$$W_E = W_1 + X_1 = W_1 + \frac{(H_{01} - S_{01} W_1)^2}{(W_1 - W_0)} \tag{2.6}$$

where X_1 is the destabilizing resonance energy. The relationship between the various energies is shown in Fig. 2.6.

In many systems self-complexing between identical molecules becomes important so that a term must be added to Ψ_N to represent the contribution from a structure where an electron has been donated from the 'acceptor' to the 'donor'. If A_1 and A_2 represent identical molecules differing only in their donor-acceptor nature,

$$\Psi_N = a\Psi(A_1, A_2) + b[\Psi(A_1^+ - A_2^-) + \Psi(A_1^- - A_2^+)] \tag{2.7}$$

and

$$W_N \cong W_0 - \frac{2(H_{01} - S_{01} W_0)^2}{(W_1 - W_0)}. \tag{2.8}$$

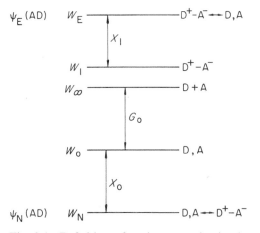

Fig. 2.6 Definition of various energies in the theory of electron donor-acceptor complexes.

[*Refs. on p. 31*]

Similarly,

$$W_E \cong W_1 + 2\frac{(H_{01} - S_{01} W_1)^2}{(W_1 - W_0)} \qquad (2.9)$$

The overlap integral S_{01} is of particular importance in pressure experiments. It can be shown that

$$S_{01} = \sqrt{2} S_{AD}(1 + S_{AD}^2)^{-1/2} \qquad (2.10)$$

where $S_{AD} = \langle \Psi_A | \Psi_D \rangle$ is the overlap integral between the highest filled molecular orbital of the donor and the lowest unfilled molecular orbital of the acceptor. Usually S_{AD} is small so S_{01} will vary linearly with S_{AD}. H_{01} will also vary linearly with S_{AD} for small S_{AD}. The overlap and orientation principle requires that the electron donor and electron acceptor moieties orient themselves such that S_{01} or S_{AD} is a maximum. Thus, in principle, it should be possible to predict the molecular geometry by maximizing S_{AD}. However, maximum overlap may be prevented if there is steric hindrance or strong localized interaction such as dipole-induced dipole, quadrupole, or hydrogen bonding. Furthermore, the symmetry of the overlap of the donor and acceptor molecules must be considered.

For the simplest case of only a single-electronic absorption characteristic of the complex, the transition arises from the $\Psi_N \to \Psi_E$ excitation. For weak interactions (small resonance energies X_0 and X_1), the transition effectively involves an intermolecular charge transfer from $\Psi_0(A, D)$ to $\Psi_1(A^- - D^+)$. The charge transfer transition energy is given by

$$h\nu_{CT} = W_E - W_N = W_1 - W_0 + \frac{(H_{01} - S_{01} W_0)^2 + (H_{01} - S_{01} W_1)^2}{(W_1 - W_0)}.$$

$$(2.11)$$

The major effect of an increase in the external pressure should be a decrease in the intermolecular separation of D and A which would cause an increase in S_{AD}. A decrease in energy with pressure. is generally observed.

A relationship between $h\nu_{CT}$ and the ionization potential of the donor I_D and the electron affinity of the acceptor E_A may be derived if it is noted that $W_0 = W_\infty - G_0$ and $W_1 = W_\infty + I_D - E_A - G_1$ where W_∞ is the energy of the separated components, G_0 is the energy of the no bond physical interactions, and G_1 is the coulombic attractive energy of the A^- and D^+ ions. Then

$$h\nu_{CT} = I_D - (E_A + G_1 - G_0) + \frac{\beta_0^2 + \beta_1^2}{I_D - (E_A + G_1 - G_0)} \qquad (2.12)$$

where $\beta_0 = (H_{01} - S_{01} W_0)$ and $\beta_1 = (H_{01} - S_{01} W_1)$.

[*Refs. on p. 31*]

If the donor-acceptor interactions are too strong for the perturbation theory to apply, the variation method must be used to determine the resonance energies. This gives

$$hv_{CT} = \frac{(W_1 - W_0)}{(1 - S_{01}^2)} \left[1 + \frac{4\beta_0 \beta_1}{(W_1 - W_0)^2} \right]^{1/2}.$$ (2.13)

As we indicated in an earlier section, the integrated intensity of an absorption band is given by the oscillator strength

$$f = k v_{max} \, \mu_{EN}^2$$ (2.14)

where k is a constant, v_{max} is the frequency at the maximum intensity of the band, and μ_{EN} is the transition dipole given by

$$\mu_{EN} = -e \int \Psi_E \sum_i \mathbf{r}_i \, \Psi_N \, d\tau$$ (2.15)

where $\mathbf{r}_i$ is the vector distance of the ith electron from an arbitrary origin. It may be expressed as

$$\mu_{EN} = a^{\dagger}b(\mu_1 - \mu_0) + (aa^{\dagger} - bb^{\dagger})(\mu_{01} - \mu_0 S_{01})$$ (2.16)

where μ_0 and μ_1 are the dipole moments of the no-bond and dative structures. With application of pressure the resonance interaction should increase and there will be an increase in the value of μ_{EN}. In view of the fact that pressure tends to decrease v_{max} it is clear that the intensity of the charge transfer bond involves a balance of the two effects.

Varying degrees of sophistication have been used in the MO analyses of intermolecular charge transfer transitions. These include the simple treatment by Dewar [23], the more detailed analyses by Murrell, [24, 25] the semiempirical linear combination of molecular orbitals approach by Flurry [26], and the treatment of Iwata *et al.* [27] which includes configuration interaction among ground, charge transfer, and locally excited states.

The third approach involves the free electron model for conjugated molecules which treats the π electron system as a free electron gas which moves in the potential field of the molecule. The concept has been developed for charge transfer complexes in the one-dimensional case by Shuler [28] and more recently by Boeyens [29]. The one-dimensional double-minimum potential well representing the electron donor-acceptor interaction is shown in Fig. 2.7. The potential barrier of height V_{AD} and width d represents an inverse measure of the delocalization of the electrons between D and A. V_{AD} should be proportional to $(I_D - E_A)$

[*Refs. on p. 31*]

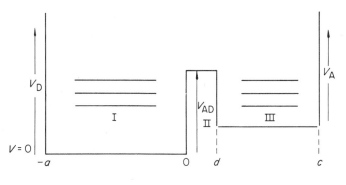

FREE ELECTRON MODEL

Fig. 2.7 Free electron model of electron donor-acceptor complexes – one dimensional potential.

and d is related to the intermolecular separation. The solid lines within each minimum represent the energy levels of the donor and acceptor molecules; the levels split when the interaction is considered. The splitting can be shown to increase with decreasing barrier height and with increasing height of the energy levels. For the case where the two potential minima have the same value, the Schrödinger equation may be solved in regions I, II and III to yield

$$\frac{\beta}{\alpha} \tan(\alpha a) + \frac{\alpha}{\beta} \cot(\alpha a) + 2 \coth(\beta d) = 0 \qquad (2.17)$$

where $\alpha = (\kappa E_{AD})^{1/2}$, $\beta = [\kappa(V_{AD} - E_{AD})]^{1/2}$, $\kappa = 2m/h^2$ and E_{AD} is an energy level of the complex. For $\beta d \geq 2$, which is true for typical values of β and d, Equation (2.17) reduces to

$$\pm \frac{1}{a(\kappa V_{AD})^{1/2}} = \frac{\sin(\alpha a)}{\alpha a}. \qquad (2.18)$$

This indicates increased splitting for decreasing V_{AD} and increasing E_{AD}. This model is interesting for pressure experiments on charge transfer complexes as the decrease of the intermolecular separation can be directly related to d and hence to the probability of tunneling through the potential barrier.

2.4 Band structure of solids

Many electronic properties of solids, especially electron transport phenomena, are best described in terms of the band theory of solids, which is developed in every text on solid state physics. Various sophisticated

[*Refs. on p. 31*]

measurements are used to elucidate the details of band structure. We shall be concerned here only with the use of optical absorption and electrical resistance to measure the relative shift of energy levels and to identify electronic transitions.

For an isolated atom, ion, or molecule the electronic energy levels consist of a discrete ground state, a series of discrete excited states, and, at still higher energies, a continuum which corresponds to ionization. As an array of atoms is brought closer together, the electronic wave functions may interact significantly. Under these circumstances, one can best describe the electronic states in terms of bands of closely spaced allowed levels separated by gaps of 'forbidden' energies. This technique is especially effective in describing electron transport in metals and semi-conductors, as well as some properties of ionic crystals. In many molecular crystals at low pressure, the band widths may be of the order of a few kT and transport may involve another mechanism. The shapes and symmetries of the energy states are controlled by the crystal symmetry and vary from place to place in the Brillouin zone. Nevertheless, it is frequently possible to relate the states to those of the parent atoms, and one can speak of an s band, or a d band, or a band of mixed s-p-d character with some significance, if not with precision.

In this picture a partially filled band should give metallic properties. (But see the discussion of the 'Mott transition' below.) If the highest band containing electrons (the valence band) is filled, one may have a metal, a semimetal, a semiconductor, or insulator depending on its location *vis-à-vis* the lowest empty band.

The electronic energy is a function of the propagation vector (κ) of the wave function. One can plot this energy in a space generated by the components of this vector. The periodicity of the atoms in real space introduces a periodicity in κ space. The 'unit cell' of this κ space structure is known as the Brillouin zone.

Fig. 2.8 represents some of the possible types of electronic events which one might observe. The cross-hatched areas represent filled states. In Fig. 2.8(a) one sees a direct transition ($\Delta\kappa = 0$) from the top of the valence band to the bottom of the conduction band. This is the normal allowed transition. At temperatures above absolute zero the atoms of the lattice are vibrating. The propagation vector of a phonon may add to or subtract from the propagation vector of the ground state and give indirect transitions where $\Delta\kappa \neq 0$ as shown in Figs. 2-8(b) and 2.8(c). There may exist bound excited states below the conduction band as shown in Fig. 2.8(d). These can be detected at atmospheric pressure by lack of photo-

[*Refs. on p. 31*]

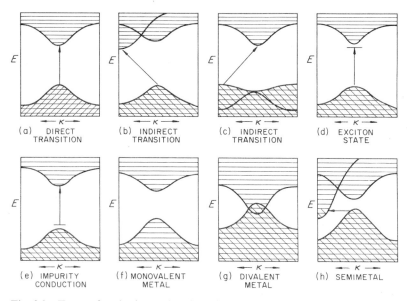

Fig. 2.8   Types of excitations – band model.

conductivity in an optically excited crystal. Such experiments have not yet been performed at very high pressure. Impurity atoms with energy levels above the top of the valence band may furnish electrons to the conduction band as seen in Fig. 2.8(e).

Figs. 2.8(f) and 2.8(g) represent the band situation in simple monovalent and divalent metals where one would expect resistance to increase linearly with increasing temperature. Fig. 2.8(h) shows a more complex type of metal, a semimetal, whose resistivity may not be linear in temperature. For the purpose of discussion of experimental results in this part of the review, a semiconductor (or insulator) is defined as a material whose resistance decreases exponentially with increasing temperature while a metal is a material whose resistance increases with temperature, whether or not the increase is linear.

The filling of the levels is given by Fermi Dirac statistics. The energy boundary between filled and empty levels is known as the Fermi level. It corresponds to the chemical potential of the system. For simple 'free electron' metals (Figs. 2.8(f-h)), it is a sphere. For more complex metals or semimetals, it may intersect the Brillouin zone boundary in a complex way, leaving holes and pockets of electrons. For intrinsic semiconductors, it lies half-way between the top of the valence band and the bottom of the conduction band.

[*Refs. on p. 31*]

With pressure the general tendency is for energy bands to broaden. In semiconductors this reduces the energy gap, and it can ultimately lead to overlap of the highest filled band with the empty conduction band, and thus to metallic behavior. It is also possible that there will be a relative shift of the center of gravity of one band with respect to the other, which can either decrease or increase the gap. Optical absorption measurements are useful in studying these events. Since the optical transition from the top of the valence band to the conduction band is usually allowed, the intense absorption at the appropriate wavelength measures the gap.

Electrical resistivity measurements also reveal information about electronic structure. One can write the resistivity:

$$\rho = \frac{1}{n\mu e} \qquad (2.19)$$

where n is the number of carriers, μ their mobility and e their charge. Electrical conduction in semiconductors is generally carrier-limited. The number of carriers is given by

$$n = n_0 \exp\left(-\frac{E_g}{2RT}\right) \qquad (2.20)$$

for intrinsic semiconductors, where E_g is the gap between the conduction and valence band.

Typically, for a metal, electron transport is mobility-limited. At ordinary temperatures if the mobility is limited by lattice scattering, the resistivity increases linearly with temperature. Since compression tends to reduce the amplitude of lattice vibrations, one anticipates a moderate decrease of resistance with increasing pressure. This is frequently observed for simple metals. Lawson [30] has reviewed the effect of pressure on metallic resistance in some detail.

While resistance measurements at high pressure are common, for a variety of reasons it is difficult to establish true resistivities in the apparatus used for very high pressure measurements. One can, however, make semiquantitative measurements of changes of the energy gaps in semiconductors with pressure, and locate insulator-metal transitions. In metals one can frequently locate first-order phase transitions. Of more interest here is the possibility of locating electronic transitions. There is frequently a sharp increase in resistance due to interband scattering near an electronic transition. This, in fact, is at present the usual experimental way of establishing the existence of this phenomenon in metals. Ordinary constitutive first-order phase transitions become very sluggish at tempera-

[*Refs. on p. 31*]

tures below room temperature as they are diffusion-controlled. This is significantly less true in electronic transitions – a factor which helps in their identification.

In the picture presented so far, materials with completely filled or completely empty bands are insulators or semiconductors (unless there is overlap of a filled with an empty band), while materials with partially filled bands are metals. However, it has been observed that many transition metal oxides are insulators or semiconductors, although some exhibit metallic conductivity. These materials have been the subject of very extensive experimental and theoretical investigation. Adler [31] presents an extensive review. Mott [32] has developed a theory which states that there is a critical electron density which a material should be an insulator and above which it is a metal. There have been extensive experimental studies in both doped semiconductors and transition metal oxides. It is not at present clear that a true Mott transition is experimentally accessible [33], but the experimental investigations, particularly on mixed transition metal oxides, have expanded our understanding of insulator-metal transitions and will be discussed briefly in Chapter 7.

References

1. A. STREITWEISER, *Molecular Orbital Theory for Organic Chemists*, John Wiley & Sons, New York (1961).
2. L. SALEM, *The Molecular Orbital Theory of Conjugated Systems*, W. A. Benjamin, New York (1966).
3. J. N. MURRELL, *The Theory of the Electronic Spectra of Organic Molecules*, Methuen, London (1963).
4. J. A. POPLE and D. L. BEVERIDGE, *Approximate Molecular Orbital Theory*, McGraw-Hill, New York (1970).
5. M. TINKHAM, *Group Theory and Quantum Mechanics*, McGraw-Hill, New York (1964).
6. R. S. MULLIKEN, C. A. RIEKE, D. ORLOFF and H. ORLOFF, *J. Chem. Phys.*, **17** 1248 (1949).
7. H. BETHE, *Ann. Physik*, **3** 133 (1929).
8. C. J. BALLHAUSEN, *Introduction to Ligand Field Theory*, McGraw-Hill, New York (1962).
9. B. N. FIGGIS, *Introduction to Ligand Fields*, Interscience, New York (1966).
10. E. U. CONDON and G. H. SHORTLEY, *The Theory of Atomic Spectra*, Cambridge University Press, London (1964).
11. G. RACAH, *Phys. Rev.*, **62** 438 (1942).
12. G. RACAH, *Phys. Rev.*, **63** 367 (1943).
13. Y. TANABE and S. SUGANO, *J. Phys. Soc. Japan*, **9** 753 (1954).
14. Y. TANABE and S. SUGANO, *J. Phys. Soc. Japan*, **9** 766 (1954).
15. J. S. GRIFFITH, *The Theory of Transition Metal Ions*, Cambridge University Press, London (1964).

16. J. S. GRIFFITH, *J. Inorg. Nucl. Chem.* **2** 229 (1956).
17. C. K. JORGENSEN, *Progr. Inorg. Chem.*, **4** 73 (1962).
18. R. S. MULLIKEN, *J. Am. Chem. Soc.*, **74** 811 (1952).
19. R. S. MULLIKEN, *J. Phys. Chem.*, **56** 801 (1952).
20. R. S. MULLIKEN, *J. Chim. Phys.*, **61** 20 (1964).
21. R. S. MULLIKEN and W. B. PERSON, *Molecular Complexes*, Wiley-Interscience, New York (1969).
22. R. FOSTER, *Organic Charge Transfer Complexes*, Academic Press, London (1969).
23. M. L. S. DEWAR and A. R. LEPLEY, *J. Am. Chem. Soc.*, **83** 4560 (1961).
24. J. N. MURRELL, *J. Am. Chem. Soc.*, **81** 5037 (1959).
25. J. N. MURRELL, *Quart. Rev.* (*London*), **15** 191 (1961).
26. R. L. FLURRY, JR., *J. Phys. Chem.* (*Ithaca*), **69** 1927 (1965).
27. S. IWATA, J. TANAKA and S. NAGAKURA, *J. Am. Chem. Soc.*, **88** 894 (1966).
28. K. E. SHULER, *J. Chem. Phys.*, **20** 1865 (1952).
29. J. C. A. BOEYENS, *J. Phys. Chem.*, **71** 2969 (1967).
30. A. W. LAWSON, *Prog. Metal Phys.* **6** 1 (1956).
31. D. ADLER in *Solid State Physics*, Vol. 21, edited by F. SEITZ, D. TURNBULL and H. EHRENREICH, Academic Press, New York (1968).
32. N. F. MOTT, *Proc. Phys. Soc.* (*London*) **A62** 416 (1949).
33. N. F. MOTT, *Comments on Solid State Physics II*, 183 (1970).

CHAPTER THREE

Thermal Versus Optical Transitions

A common technique for measuring the difference in energy between the ground state and an excited electronic state is by optical absorption which, of course, gives a measure of this difference, subject to definite restrictions. On the other hand, the process of electron transfer to establish a new ground state at high pressure is a thermal one. As we shall find it useful to relate these processes, it is important to understand their differences. We shall find that in many cases the thermal energy is apparently much smaller than the energy observed optically. We shall show that this is reasonable, and present an approximate relationship between them.

First we note that a large difference in energy between optical and thermal transitions has been observed for a number of processes at one atmosphere. The color centers induced in alkali halides by X-irradiation or by excess alkali halide characteristically absorb in the visible or ultra-violet regions of the spectrum (2-4 eV) [1]. Yet they can be bleached thermally at moderate temperatures – in some cases as low as 100 °K (300 °K $\sim$ 0.025 eV).

In redox reactions in aqueous solutions [2] one measures electron transfer between a given metallic ion in two oxidation states (e.g. $Ti^{3+} \rightleftarrows Ti^{4+}$, $Cr^{3+} \rightleftarrows Cr^{2+}$, etc.). The thermal energy associated with the electron transfer is, of course, zero, since the ground states are the same, but these solutions typically absorb in the region 2-3 eV.

There are a number of reasons for the difference in energy associated with the two types of transition; two of these can be observed in the schematic configuration co-ordinate diagram of Fig. 3.1. Here the

33   [Refs. on p. 43]

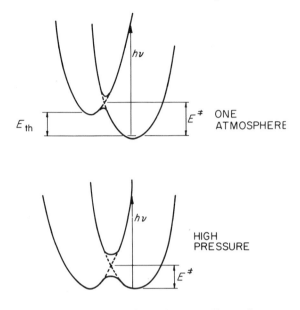

Fig. 3.1 Schematic configuration co-ordinate diagram.

potential energies of the ground and excited states are plotted against some characteristic displacement of the system.

In the first place, optical transitions are subject to the Franck-Condon principle, while thermal transitions are not. That is to say, optical transitions occur rapidly compared with nuclear motions, so that they are represented vertically (or nearly so, subject to the uncertainty principle) on configuration co-ordinate diagrams. Thermal transitions are much slower and can take advantage of nuclear rearrangement.

In the second place, the mixing of states by configuration interaction is a common phenomenon. This results from partial relaxation of the Born-Oppenheimer condition due to spin-orbital coupling. In a solid of the complexity of those considered in much of this work, there will always be an appropriate vibration to mix two states of almost any symmetry. As can be seen from Fig. 3.1, configuration interaction can increase significantly the difference between the energies associated with optical and thermal transitions.

There are two other important factors not illustrated in Fig. 3.1. Optical processes are subject to parity selection rules ($g \rightarrow u$ or $u \rightarrow g$). In the time scale associated with these thermal transitions any selection rules formally associated with them are relaxed. Since there can be energy differences of the order of 1 eV or more between the highest occupied

[*Refs. on p. 43*]

levels of g and u parity, this factor can make a significant contribution to the difference under consideration.

Finally, one must recognize that the diagram in Fig. 3.1 is grossly oversimplified in that it indicates only a single configuration co-ordinate. In a pressure-induced thermal process the pressure selects the volume of the system as the appropriate configuration co-ordinate. Optical processes in general involve more configuration co-ordinates, which can serve also to increase the difference in energy involved in the two processes. This, in fact, is a limiting factor in the analysis presented below. Electronic degeneracy has also been ignored.

The radiationless transition of the optically excited electron to the bottom of the excited state potential well must be a multiphonon process. Between the molecular vibrations and the lattice vibrations there would appear always to be ample modes available to dissipate the energy.

There have been a number of theoretical treatments of one aspect or another of electron transfer processes by Marcus, [3, 4] Hush, [5, 6] and by Henry and Slichter [7]. We extend these analyses to include the effect of pressure, with potential wells of differing force constants [8].

A schematic diagram is presented in Fig. 3.2. We assume a single configuration co-ordinate and harmonic potential wells with different force constants ω and ω'. At 1 atm the potential well of the excited state is displaced from that of the ground state by a quantity Δ. The energy difference between the bottom of the two wells is E_{th}; at 1 atm we label it E_0. The quantities E^+ and E^- are discussed later. Upon application of pressure the potential energy of the ground state can be written

$$V_g = \tfrac{1}{2}\omega^2 Q^2 + pQ. \tag{3.1}$$

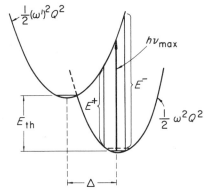

Fig. 3.2 Schematics of optical and thermal excitations.

[*Refs. on p. 43*]

In the second term on the right one assumes that the pressure works against an area A which is independent of pressure so that it can be incorporated into p or Q. Similarly, for the excited state one can write

$$V_e = \tfrac{1}{2}(\omega')^2(Q - \Delta)^2 + pQ + E_0. \tag{3.2}$$

At the potential minimum

$$\frac{\partial V_g}{\partial Q} = 0 = \omega^2 Q + p \tag{3.3}$$

so

$$Q = \frac{-p}{\omega^2} \tag{3.4}$$

and

$$\frac{\partial V_e}{\partial Q} = 0 = (\omega')^2(Q - \Delta) + p \tag{3.5}$$

so

$$Q - \Delta = \frac{-p}{(\omega')^2}. \tag{3.6}$$

We assume throughout this part of the analysis that ω and ω' are independent of pressure. Then,

$$(V_g)_{min} = -\frac{1}{2}\frac{p^2}{\omega^2} \tag{3.7a}$$

and

$$(V_e)_{min} = p\Delta - \frac{p^2}{2(\omega')^2} + E_0 \tag{3.7b}$$

If we let $Q' = Q + p/\omega^2$, $Q' = 0$ is then the bottom of the new ground state. Substituting in (3.1) and (3.2) we get

$$V_g = (V_g)_{min} + \tfrac{1}{2}\omega^2(Q')^2 \tag{3.8a}$$

and

$$V_e = (V_e)_{min} + \tfrac{1}{2}(\omega')^2(Q' - \Delta')^2 \tag{3.8b}$$

where

$$\Delta' \equiv \Delta + p\left[\frac{1}{\omega^2} - \frac{1}{(\omega')^2}\right]. \tag{3.9}$$

[*Refs. on p. 43*]

Equations (3.8a and b) show that we can eliminate the explicit pressure dependence by going to a new co-ordinate Q' and considering harmonic wells of altered energies and relative displacements in co-ordinate, but of unaltered spring constants.

Then E_{th}, the difference between the bottoms of the potential wells, is

$$E_{th} = E_0 + p\Delta + \frac{p^2}{2}\left[\frac{1}{\omega^2} - \frac{1}{(\omega')^2}\right]. \qquad (3.10)$$

At zero pressure

$$h\nu_{max} = E_0 + \tfrac{1}{2}(\omega')^2\Delta^2 \qquad (3.11)$$

so, at pressure p

$$h\nu_{max} = E_{th} + \tfrac{1}{2}(\omega')^2(\Delta')^2 \qquad (3.12)$$

$$= E_0 + p\Delta + \frac{p^2}{2}\left[\frac{1}{\omega^2} - \frac{1}{(\omega')^2}\right] + \frac{(\omega')^2}{2}\left\{\Delta + p\left[\frac{1}{\omega^2} - \frac{1}{(\omega')^2}\right]\right\}^2. \qquad (3.13)$$

From Equations (3.12) and (3.13) we see that there are two components which affect the shift of the optical absorption peak maximum with pressure. In addition to the vertical displacement of the potential wells (the change in E_{th}), there is also the possibility of a horizontal displacement along the configuration co-ordinate (a change from Δ to Δ'), that is to say, different compressibilities for the ground and excited states along that co-ordinate. Since these may have opposing effects on ν_{max}, a relatively small change in location of the optical absorption peak maximum may be consistent with a significant change in E_{th}.

We shall now develop a formula for the peak width. We shall first define a width δE which is proportional to but not equal to the Gaussian half width $\delta E_{1/2}$, and then transfer to the latter value.

For a harmonic oscillator one can express the energy at the terminal points of the oscillation:

$$\tfrac{1}{2}\omega^2 Q^2 = \tfrac{1}{2}kT \qquad (3.14)$$

$$Q\pm = \pm\left(\frac{kT}{\omega^2}\right)^{1/2}. \qquad (3.15)$$

Note $|Q^+| = |Q^-|$. Then:

$$E^+ = E_0 + \tfrac{1}{2}(\omega')^2(Q^+ - \Delta)^2 - \tfrac{1}{2}\omega^2(Q^+)^2 \qquad (3.16)$$

$$E^- = E_0 + \tfrac{1}{2}(\omega')^2(Q^- - \Delta)^2 - \tfrac{1}{2}\omega^2(Q^-)^2. \qquad (3.17)$$

[*Refs. on p. 43*]

Then, for zero pressure

$$(\delta E)_{p=0} = \left| \frac{-2(\omega')^2 \Delta}{\omega} \right| (kT)^{1/2}. \tag{3.18}$$

At $p \neq 0$ we simply replace Δ by Δ', so

$$(\delta E)_p = \left| \frac{-2(\omega')^2}{\omega} \left\{ \Delta + p \left[\frac{1}{\omega^2} - \frac{1}{(\omega')^2} \right] \right\} \right| (kT)^{1/2}. \tag{3.19}$$

This contains the important implication that the peak width changes with pressure only if $\omega' \neq \omega$, i.e., if the compressibilities of the two states are different.

At $p = 0$,

$$h\nu_{max} = E_0 + \frac{1}{8} \frac{(\delta E)^2}{kT} \frac{\omega^2}{(\omega')^2}. \tag{3.20}$$

This equation will, however, hold at any pressure since Δ, and thus Δ', has been eliminated. Following Hush [6], one can transform from the width δE to the half width of a Gaussian peak $\delta E_{1/2}$, then:

$$h\nu_{max} = E_{th} + \frac{1}{16 \ln 2} \frac{(\delta E_{1/2})^2}{kT} \frac{\omega^2}{(\omega')^2}. \tag{3.21}$$

At 25 °C, if one expresses $\delta E_{1/2}$ in eV, one obtains

$$h\nu_{max} = E_{th} + 3.6(\delta E_{1/2})^2 \frac{\omega^2}{(\omega')^2}. \tag{3.22}$$

This constitutes a relationship between the observed maximum for optical absorption, the thermal energy, and the half width of the peak observed. A number of limitations are discussed below. Nevertheless, some interesting comparisons between systems can be made. Unfortunately, most of the optical absorption measurements made so far at high pressure are not sufficiently accurate to use these equations in their full strength. On using Equation (3.22) relating $h\nu_{max}$ and $\delta E_{1/2}$, it is usually necessary to make a rough estimate of the ratio $(\omega/\omega')^2$. For several systems where calculations can be made, the factor is in the range 0.9 to 0.98 even when the half width changes by 50% in 100 kbar. Thus for rough calculations of the pressure where $E_{th} = 0$, even the assumption $(\omega'/\omega)^2 = 1$, which is not strictly consistent with a pressure-dependent peak half width (see Equation 3.19), gives a reasonable approximation.

[*Refs. on p. 43*]

From the above equations one can derive a number of expressions which should be subject to evaluation from sufficiently precise data:

$$\left(\frac{\partial h\nu_{max}}{\partial p}\right) = \frac{(\omega')^2}{\omega^2}\left\{\Delta + p\left[\frac{1}{\omega^2} - \frac{1}{(\omega')^2}\right]\right\} \tag{3.23}$$

$$\left(\frac{\partial h\nu_{max}}{\partial p}\right)_{(p=0)} = \frac{(\omega')^2}{\omega^2}\Delta \tag{3.24}$$

$$\left(\frac{1}{\delta E_{1/2}}\right)_0\left(\frac{\partial \delta E_{1/2}}{\partial p}\right) = \frac{(1/\omega)^2 - (1/\omega')^2}{\Delta}. \tag{3.25}$$

There are a number of limitations on the above results, some of which can be rather easily removed. It is possible that ω and ω' are pressure-dependent. As a first assumption it is possible that terms like $(\omega')^2/\omega^2$ vary slowly with pressure compared with $\tau = [1/\omega^2 - 1/(\omega')^2]$. Then

$$\left(\frac{\partial h\nu_{max}}{\partial p}\right) = \frac{(\omega')^2}{\omega^2}(\Delta + p\tau) + p\frac{\partial \tau}{\partial p}\left[\frac{p}{2} + (\omega')^2(\Delta + p\tau)\right] \tag{3.26}$$

$$\frac{1}{(\delta E_{1/2})_0}\left(\frac{\partial \delta E_{1/2}}{\partial p}\right) = \frac{\tau + p\,\partial\tau/\partial p}{\Delta}. \tag{3.27}$$

It requires, of course, very accurate data to differentiate these equations from Equations (3.23) and (3.25) above.

A serious approximation involved in this calculation is the assumption of a single configuration co-ordinate. As indicated earlier optical transitions generally involve co-ordinates other than the volume. More generally one must recognize that the pressure couples to a number of modes so that the extra potential energy arising from pressure, V_p becomes

$$V_p = p\sum_i c_i Q_i. \tag{3.28}$$

We find then

$$\Delta'_i = \Delta_i + pc_i\left[\frac{1}{\omega_i^2} - \frac{1}{(\omega'_i)^2}\right] \tag{3.29a}$$

$$(V_g)_{min} = -\frac{1}{2}p^2\sum_i \frac{c_i}{\omega_i^2} \tag{3.29b}$$

and so forth (i.e., p is replaced by pc_i, and the quantities Δ, Δ', ω, ω' acquire subscripts i)

$$h\nu_{max} = E_{th} + \frac{1}{2}\sum_i (\omega'_i)^2(\Delta'_i)^2 = E_{th} + \frac{1}{16kT\ln 2}\sum_i (\delta E_{i1/2})^2\frac{\omega_i^2}{(\omega'_i)^2}. \tag{3.30}$$

[*Refs. on p. 43*]

If we define an average ratio of the squares of the spring constants

$$\left\langle \frac{\omega_i^2}{(\omega_i')^2} \right\rangle \equiv \frac{\sum_i (\omega_i^2/(\omega_i')^2)(\delta E_{i1/2})^2}{(\delta E_{1/2})^2} \tag{3.31}$$

where $(\delta E_{1/2})$ is the total Gaussian half width of the optical line, we get the relation

$$h\nu_{max} = E_{th} + \frac{(\delta E_{1/2})^2}{16kT \ln 2} \left\langle \frac{\omega_i^2}{(\omega_i')^2} \right\rangle. \tag{3.32}$$

If we represent the physical system by only a few normal modes, and assume pressure couples to only one, Equation (3.23) and (3.24) remain the same except that subscripts 1 are added to ω, ω', Δ, and Equation (3.25) is replaced by

$$\left| \frac{\partial(\delta E_{1/2})}{\partial p} \right| = \left| \frac{2(\omega')^2}{\omega_1} \left(\frac{1}{\omega_1^2} - \frac{1}{(\omega_1')^2} \right) \right| (kT)^{1/2}. \tag{3.33}$$

Again we note that the difference between ω_1^2 and $(\omega_1')^2$ is responsible for the pressure-dependence of the width.

If we assume a value of $\langle \omega_i^2/(\omega_i')^2 \rangle$ we can apply Equation (3.32) to compute E_{th} from measured $h\nu_{max}$ and $(\delta E_{1/2})^2$.

The effect of shear is sometimes of interest in high pressure work. In the context of the analysis presented here, shear could affect the magnitude of the variables Δ, ω, ω', and E_{th}, defined in Fig. 3.2. Large shear distortions could remove degeneracies among normal modes and increase the importance of some configuration co-ordinates which are not significant for thermal processes under purely hydrostatic conditions. It could also intensify or otherwise modify configuration interaction. The magnitude of the energy barrier between the two ground states, discussed below, could be affected by strong shear forces. In principle, shear effects as well as hydrostatic effects are scientifically interesting, but the presence of a large and unresolved component of shear associated with a quasi-hydrostatic experiment could complicate the analysis.

It is, at times, of use to calculate the energy $E^{\ddagger}$, exhibited in Fig. 3.1 at the intersection between the potential wells, as an approximation to the

[Refs. on p. 43]

energy barrier between the two ground states. For the case $\omega = \omega'$, analyses by Marcus [3, 4] and Hush [5, 6] give for this energy:

$$E^{\ddagger} = \frac{(h\nu_{max})^2}{4(h\nu_{max} - E_{th})}. \tag{3.34}$$

The analysis for $\omega \neq \omega'$ gives a much more complex relationship which is not of quantitative use in view of the limitations discussed below. In the first place, configuration interaction, as illustrated qualitatively in Fig. 3.1, ensures that the energy barrier will be less than $E^{\ddagger}$. Hush [6] has presented an analysis for the system Cr^{3+}-Cr^{2+} with H_2O ligands. He showed that the difference between $E^{\ddagger}$ and the top of the barrier increases rapidly with decreasing distance between oxygens, and should therefore increase with pressure.

In the second place, tunnelling through the barrier could reduce the effective barrier height, although for the relatively heavy atoms typically involved in the high pressure studies discussed here, the probability of tunnelling is not high. Nevertheless, Equation (3.34) may be useful as a first approximation for comparing the relative change in the barrier with pressure for a series of related compounds.

It should be noted that Equations (3.18), (3.23), (3.24) and (3.25) [or (3.33)] relate the coefficients ω, ω', and Δ which describe the potential energy configuration of the ground and excited states. They can therefore be evaluated with some redundancy, even from data over a moderate pressure range. From higher pressure optical absorption data one may be able to calculate the change of ω and ω' with pressure. Because of the simplicity of the model the absolute values of these variables for individual compounds may have limited meaning, but their systematic evaluation for a series of related molecules should give valuable information concerning energy configurations, both at 1 atm and at high pressure.

The analysis presented in this chapter applies to optical absorption. An analogous derivation can be made for emission (fluorescence or phosphorescence). For the case of a single configuration coordinate one obtains:

$$h\nu_{max} = E_0 + p\Delta + \frac{p^2}{2}\tau - \frac{1}{2}\omega^2[\Delta + p\tau]^2 \tag{3.35}$$

$$= h\nu_0 + \frac{\omega^2}{(\omega')^2}p\left[\Delta + \frac{p}{2}\tau\right] \tag{3.36}$$

[*Refs. on p. 43*]

and

$$\delta E = \left| 2 \frac{\omega^2}{\omega'} (kT)^{1/2} [\Delta + p\tau] \right| \tag{3.37}$$

or

$$\delta E_{1/2} = \left| \frac{\omega^2}{\omega' \ln 2} (kT)^{1/2} [\Delta + p\tau] \right|. \tag{3.38}$$

The above equations are analogous to (3.13) and (3.19) for absorption.

Phosphorescence usually involves a transition from a triplet state to the ground singlet state. Equations (3.36) and (3.38) together with equations analogous to (3.23)–(3.25), can be used to evaluate the force constant (say ω'') of the triplet state and its displacement (say Δ'') with respect to the (singlet) ground state along the configuration coordinate.

For fluorescence the initial and final electronic states are the inverse of these involved in absorption. If we assume no rearrangement of nuclear coordinates between absorption and fluorescence, several interesting results are obtained. At zero pressure:

$$h\nu_{(abs.-emiss.)} = \frac{(\omega')^2 + \omega^2}{2} \Delta^2, \tag{3.39}$$

a quantity which is positive definite, as it should be.

At any pressure p:

$$h\nu_{(abs.-emiss.)} = \frac{(\omega')^2 + \omega^2}{2} (\Delta + p\tau)^2 \tag{3.40}$$

Thus, if Δ and τ have the same sign the emission peak will shift with pressure more than the absorption peak. If the signs are opposite, they will draw together with increasing pressure.

Further:

$$\frac{(h\nu_1 - h\nu_2)_{abs.}}{(h\nu_1 - h\nu_2)_{emiss.}} = \left(\frac{\omega'}{\omega} \right)^4 \tag{3.41}$$

Here the 1 and 2 refer to two different pressures.

At any pressure:

$$\frac{(\delta E_{1/2})_{abs.}}{(\delta E_{1/2})_{emiss.}} = \left(\frac{\omega'}{\omega} \right)^3 \tag{3.42}$$

A measurement of both absorption and fluorescence peaks as a function of pressure for a related series of compounds would be very helpful in characterizing electronic states.

[*Refs. on p. 43*]

References

1. J. H. SHULMAN and W. D. COMPTON, *Color Centers in Solids*, McMillan, New York, (1962).
2. H. TAUBE, *Adv. Inorg. Chem. Radiochem.*, **1** 1 (1959).
3. R. A. MARCUS, *Ann. Rev. Phys. Chem.*, **15** 155 (1964).
4. R. A. MARCUS, *J. Chem. Phys.*, **43** 1261 (1965).
5. N. S. HUSH, *Prog. Inorg. Chem.* Vol. 8, p. 357, edited by F. A. Cotton Interscience, New York, (1967).
6. N. S. HUSH, *Electrochim. Acta.*, **13** 1005 (1968).
7. C. H. HENRY and C. P. SLICHTER, in *Physics of Color Centers*, edited by W. B. Fowler Academic Press, New York, (1968).
8. H. G. DRICKAMER, C. W. FRANK, and C. P. SLICHTER, *Proc. Nat. Acad. Sci.*, **69** 933 (1972).

CHAPTER FOUR
Phenomenological Description of
Continuous Electronic Transitions

Many electronic transitions in metals occur at some fixed pressure at a given temperature and are accompanied by a volume discontinuity. On the other hand, some electronic transitions, including those in compounds of iron involving changes in spin state or oxidation state, occur continuously over a range of pressure. In fact, for the reduction of ferric to ferrous iron, one can frequently represent the conversion over a considerable range of pressure by the relationship:

$$K = \frac{C_{\text{II}}}{C_{\text{III}}} = Ap^M \qquad (4.1)$$

where C_{II} and C_{III} are the fractions of ferrous and ferric ions, p is the pressure and A and M are constants. For some transformations the conversion at first increases with pressure and then levels or may actually go through a maximum.

It is easy to exaggerate the difference between the discontinuous and continuous behavior. As discussed in Chapter 7, for some transitions the magnitude of the discontinuity decreases with increasing temperature so that there is a critical point above which there is no discontinuity in physical properties. Phenomena associated with the electronic transition may also extend above the melting temperature and persist over a range of pressure and temperature in the liquid, with no clear discontinuity in the properties at a transition point. On the other hand, it is a common phenomenon for a reactant and product to exist together in equilibrium in the liquid phase over a range of temperature and pressure, with the relative concentrations depending on the reaction conditions. However,

[*Refs. on p. 59*]   44

for this case there can also be a critical temperature below which a product may precipitate, forcing the reaction to completion. Thus, the 'continuous' or 'discontinuous' nature of the process may depend on the external conditions.

A phenomenological description has been developed [1], primarily to describe electronic transitions in iron compounds as a function of temperature and pressure. It is analogous in many ways to the regular solution theory of chemistry or to molecular field theories of magnetism. It predicts all of the general features observed in high pressure iron chemistry. Unfortunately, the experimental data are not sufficiently precise to permit quantitative calculations for specific systems. The theory is, however, very general and should apply to a wide variety of phenomena involving interacting systems. We therefore outline it in some detail below.

One point of caution should be made. A thermodynamic theory requires, of course, that the system be in equilibrium, and equilibrium may be very hard to establish in a reacting solid. A system may remain indefinitely in a state significantly above the lowest energy state if there is a sufficient energy barrier between states. An example is the existence of glasses instead of crystalline solids. The theory presented below, however, contains the essential features observed experimentally in a wide variety of systems.

We consider a mixture of two components with Gibbs free energy G_0 and G_1. The relative amounts at equilibrium depend on pressure and temperature.

We write the free energy for the mixture:

$$G = N_0[(1 - c)G_0(p, T) + cG_1(p, T) + \Gamma(p, T)c(1 - c)] - T\sigma_{mix} \quad (4.2)$$

where Γ is an interaction term which indicates that the free energy of formation may depend on the site fraction converted, and

$$\sigma_{mix} = k_B[N_0 \ln N_0 - N_0 c \ln N_0 c - N_0(1 - c) \ln N_0(1 - c)] \quad (4.3)$$

is the usual mixing term due to the fact that there are a variety of ways of choosing the converted sites. The symmetry around $c = 0.5$ is an artifact which could be changed by keeping higher terms or choosing a different form for the coefficient of Γ. If one makes use of the condition:

$$\left(\frac{\partial G}{\partial c}\right)_{p, T} = 0 \quad (4.4)$$

[*Refs. on p. 59*]

for equilibrium, one obtains

$$G_1 - G_0 + \Gamma(1 - 2c) + k_B T \ln\left(\frac{c}{1-c}\right) = 0 \qquad (4.5)$$

and therefore

$$\ln K = -\frac{1}{k_B T}[\Delta G(p, T) + \Gamma(p, T)(1 - 2c)]. \qquad (4.6)$$

One can then obtain results of varying degrees of sophistication by assuming various forms for ΔG and Γ. If one assumes $\Gamma = 0$, then

$$\ln K = -\frac{\Delta G(p, T)}{k_B T}. \qquad (4.7)$$

4.1 Linear elastic theory – Non-interacting systems

We consider first a case where the elastic response of the lattice is linear $\left(\text{i.e., the bulk modulus } B = -V\left(\frac{\partial p}{\partial V}\right)_T \text{ is independent of pressure}\right)$. Then

$$\ln K = -\frac{1}{k_B T}\left[\Delta G(0) + p\Delta \frac{\partial G}{\partial p}(0) + \frac{p^2}{2}\Delta \frac{\partial^2 G}{\partial p^2}(0)\right] \qquad (4.8)$$

where the symbol (0) refers to the value at zero pressure. This can be written

$$\ln K = -\frac{1}{k_B T}\left[\Delta G(0) - p\Delta V(0) + \frac{p^2}{2}\Delta \frac{V(0)}{B}\right] \qquad (4.9)$$

where the bulk modulus B is evaluated at $p = 0$.

The first term represents the contribution from the free energy difference at zero pressure, the second term is due to the fact that the two components have different volumes, and the third term is the result of differing bulk moduli. One sees that if one component has a larger volume, but also a larger compressibility (smaller B), the possibility of a maximum in ln K versus pressure exists.

If we define

$$x = \frac{p\Delta\left(\frac{V}{B}\right)}{2\Delta V} \qquad (4.10)$$

$$p\frac{\partial \ln K}{\partial p} = -\frac{1}{k_B T}\left[p\Delta V(0) - p^2\Delta \frac{V(0)}{B}\right] \qquad (4.11)$$

$$= -Ax(1 - 2x). \qquad (4.12)$$

[*Refs.* on p. 59]

In terms of solution theory, the right hand side is just $\dfrac{p\,\Delta\bar{V}}{k_B T}$. The product $x(1-2x)$ varies slowly with pressure and approximates the straight line relationship of Equation (4.1). It is not surprising that these equations give only a very qualitative approximation to experiment, as we have neglected Γ and because it is well known that in the pressure range necessary for these conversions B is definitely dependent on pressure.

4.2 Non-linear elastic theory – non-interacting systems

We assume a form for the Helmholtz free energy A:

$$A = \sum_m \frac{A_m}{V^m} \tag{4.13}$$

where the coefficients A_m depend on temperature but not volume. Later we discuss the typical values of m for the types of atomic interactions involved in transition metal complexes. Then

$$p = \sum_m \frac{mA_m}{V^{m+1}} \tag{4.14}$$

and B, the bulk modulus at zero pressure, is given by

$$B = \sum_m \frac{m^2 A_m}{V^{m+1}}. \tag{4.15}$$

Utilizing the fact that at zero pressure

$$\sum_m \frac{mA_m}{V_0^m} = 0 \tag{4.16}$$

then

$$G = \sum_m \frac{(m+1)A_m}{V^m}. \tag{4.17}$$

One can then calculate ΔG between two states at the same pressure but different volumes:

$$\Delta G = \sum_m \frac{\delta A_m}{V^m} \tag{4.18}$$

where V is the volume of the unconverted state.

[*Refs. on p. 59*]

A convenient relationship between p and V is the Murnaghan [2] equation

$$\frac{V_0}{V} = \left(1 + \frac{np}{B}\right)^{1/n}.$$
(4.19)

Defining $y = \dfrac{np}{B}$ we obtain:

$$\ln K = -\frac{1}{k_B T}\left[\sum_m \frac{\delta A_m}{V_0^m}(1 + y)^{m/n}\right]$$
(4.20)

$$= -\frac{1}{k_B T}\sum_m \Delta G_m(1 + y)^{m/n}$$
(4.21)

where

$$\Delta G_m = \frac{\delta A_m}{V_0^m}$$
(4.22)

In general there may be variety of terms corresponding to different values of m. For an approximate treatment of metal complexes we take the values $m = 1/3, 5/3, 9/3$. The first term includes primarily coulomb attraction plus other terms with small volume-dependence. The second term approximates crystal field and covalent effects, while the last term represents closed shell repulsion. Using a value of $n = 16/3$, reasonable for ionic crystals:

$$\ln K = -\frac{\Delta G_1}{k_B T}(1 + y)^{1/16}\left[1 + \frac{\Delta G_5}{\Delta G_1}(1 + y)^{1/4} + \frac{\Delta G_9}{\Delta G_1}(1 + y)^{1/2}\right].$$
(4.23)

Typically, for charge transfer ΔG_1 will be positive, ΔG_5 negative, and ΔG_9 positive. For appropriate values of the ΔG_i one can obtain roughly linear plots of $\ln K$ versus $\ln p$ over some ranges of conversion, and also maxima. Detailed plots and tables are included in the original paper. However, to approximate experiment reasonably closely one must assume a very close cancellation between ΔG_1 and ΔG_5, which would seem to be fortuitous.

4.3   Interacting centers

To the first order we can expand Γ in a power series in p

$$\Gamma(p, T) = \Gamma_0(T) + p\Gamma_1(T) + p^2\Gamma_2(T).$$
(4.24)

Here Γ_0 represents the fact that at zero pressure the conversion of a

[*Refs. on p.* 59]

given ion or molecule requires a different free energy at low concentration than at high. It may be positive or negative, depending on whether the Madelung energy associated with charge transfer or the elastic strain associated with misfit of converted molecules dominates. A positive Γ_0 corresponds to an increase of conversion, and a negative Γ_0 a decrease. As is shown in the original paper, Γ_1 is proportional to $\left(\dfrac{\Delta V}{V}\right)\left(\dfrac{\Delta B}{B}\right)$ and Γ_2 is proportional to $\left(\dfrac{\Delta B}{B}\right)^2$, where ΔV and ΔB are the differences in volume and bulk modulus between the two types of material.

One can best observe the effect of Γ_0 on conversion by considering the pressure and temperature coefficients of $\ln K$.

$$\frac{d \ln K}{d \ln p}\left[1 - \frac{\Gamma_0}{k_B T} \frac{2K}{(1 + K)^2}\right] = \left(\frac{d \ln K}{d \ln p}\right)_{\Gamma_0 = 0}. \tag{4.25}$$

If we consider the particular pressure where $K = 1$, the equation simplifies to

$$\frac{d \ln K}{d \ln p} = \left[\frac{1}{1 - \dfrac{\Gamma_0}{2k_B T}}\right]\left(\frac{d \ln K}{d \ln p}\right)_{\Gamma_0 = 0} \tag{4.26}$$

For repulsion (Γ_0 negative) there may be a substantial reduction in slope. For $\Gamma_0 = -0.1$ eV the slope is cut by a factor of 3, while for $\Gamma_0 = -0.3$ eV the slope is cut by a factor of 7. These values are still a very small fraction of ΔG_1 or ΔG_5. The general effect of Γ is to straighten out the curve of $\ln K$ versus $\ln p$ and to give slopes which correspond more closely to experimental results.

The coupling of Γ can produce co-operative effects which affect the temperature-dependence of the conversion.

$$\frac{d \ln K}{d\left(\dfrac{1}{T}\right)} = -\frac{1}{k_B} \frac{\Delta H + \left(\dfrac{1 - K}{1 + K}\right)\left[\Gamma - T \dfrac{d\Gamma}{dT}\right]}{\left[1 - \dfrac{2\Gamma}{k_B T} \dfrac{K}{(1 + K)^2}\right]} = -\frac{\Delta H_{\text{eff}}}{k_B}. \tag{4.27}$$

Thus the effective ΔH will vary with Γ and with temperature. At $K = 1$ we find

$$\Delta H_{\text{eff}} = \frac{\Delta H}{1 - \dfrac{\Gamma}{2k_B T}}. \tag{4.28}$$

[*Refs. on p. 59*]

If Γ is negative the effective ΔH is reduced by the same ratio as is the slope of $\ln K$ versus $\ln p$. If Γ is positive the denominator could go to zero giving an infinite change of $\ln K$ with temperature. We discuss this effect, along with others, in the final section.

4.4 Graphical methods

In the theory of ferromagnetism, a transcendental equation must be solved to obtain the magnetization. There is a well known graphical solution to the equation. We expect from the physical analogy to our problem that a graphical means should exist for finding c. One such approach is to note that if the two sides of Equation (4.6) are plotted separately as functions of c, the resulting intersection gives those values of c which satisfy Equation (4.6). It is convenient to rewrite equation (4.6) as follows.

$$\ln\left(\frac{1-c}{c}\right) = \frac{\Delta H + \Gamma(1-2c)}{k_B T} - \frac{\Delta S}{k_B} \qquad (4.29)$$

The left hand side vanishes at $c = 1/2$ and is an odd function of c about that point. The right hand side is a straight line of slope $-\dfrac{2\Gamma}{k_B T}$.

At $c = \frac{1}{2}$ the coefficient of the term in Γ vanishes. At $c = \left(\dfrac{1}{2} + \dfrac{\Delta H}{2\Gamma}\right)$ the term in $1/T$ vanishes. We will use these facts below.

4.4.1 Non-interacting centers $(\Gamma = 0)$

For non-interacting centers, the graphical solution of Equation (4.29) is shown in Fig. 4.1. The right hand side is horizontal (zero slope) and the solution is given by the value of c corresponding to point A. The case shown has both ΔS and ΔH positive. The point B corresponds to the solution for $T = \infty$ (or for $\Delta H = 0$), and demonstrates that the term in ΔS 'biases' the infinite temperature site fraction. It can be shown that this simple case cannot account for many of the experimental observations.

4.4.2 Interacting centers $(\Gamma \neq 0)$

From our earlier remarks, we can generate the graphical solution by drawing the straight line with slope $-2\Gamma/k_B T$ through the point $\left(\dfrac{1}{2}, \dfrac{\Delta H}{k_B T} - \dfrac{\Delta S}{k_B}\right)$. The resulting construction is shown in Fig. 4.2. Again point A represents the solution. Whether or not the slope changes depends on whether or not Γ depends on pressure.

[Refs. on p. 59]

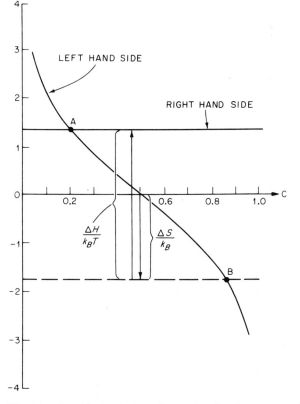

Fig. 4.1 Graphical solution of Equation (4.29) for non-interacting centers ($\Gamma = 0$).

We note that if the slope is negative (attractive interaction) and sufficiently steep, the straight line can intersect the logarithmic curve at three places. This fact is illustrated in Fig. 4.3 for the case $\Delta H = 0$, $\Delta S = 0$. These circumstances correspond to the molecular field treatment of the Ising lattice of spin $\frac{1}{2}$ in zero external magnetic field. c represents the number of spins of one orientation (say spin up). Curve 1 intersects at points A, A′, A″. Examination of $\partial^2 G/\partial c^2$ shows that points A and A″ are minima in the Gibbs free energy, hence are stable, but point A′ is a maximum, hence unstable. The symmetry of c and $1 - c$ for A and A″ respectively represents the fact that there is no preferred direction of magnetization.

Curve 3 being less steep is at a higher temperature. It intersects the logarithmic curve at only one place ($c = \frac{1}{2}$) giving just one root. Curve 2, which is tangent to the log curve, corresponds to the temperature which

[*Refs. on p. 59*]

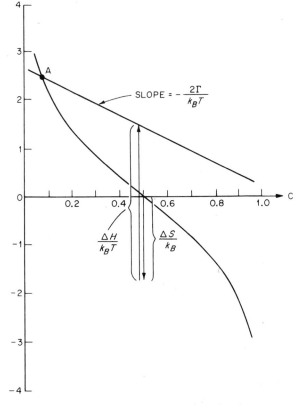

Fig. 4.2   Graphical solution of Equation (4.29) for interacting centers ($\Gamma \neq 0$).

divides the three root region of temperature from the one root one.
It therefore corresponds to the Curie temperature. In the usual theory of
solutions in which the number of moles of the two substances is a given
quantity, the temperature of curve 1 would fall in the region in which the
solution segregated into two phases, one rich in one component, the other
rich in the other.

The constructions of Fig. 4.3 were relatively easy to make since the
point through which the straight line went did not move with temperature
as a result of the fact $\Delta H = 0$. It is always possible to find a point through
which the straight line passes but which does not change with temperature
provided Γ is independent of temperature. The construction is shown in
Fig. 4.4, and consists of drawing the straight line through the point
$$\left(\frac{1}{2} + \frac{\Delta H}{2\Gamma}, \; -\Delta S/k_B \right).$$

[*Refs. on p. 59*]

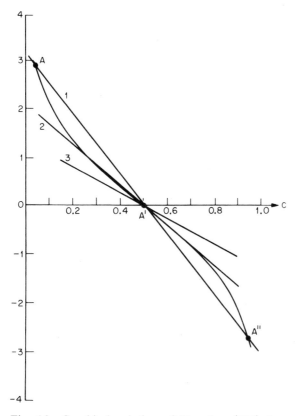

Fig. 4.3 Graphical solution of Equation (4.29) for $\Delta H = 0$, $\Delta S = 0$ showing possibility of co-operative phenomena. Curves labelled 1, 2, and 3 are at progressively higher temperatures.

Fig. 4.5 illustrates an interesting situation. Curves 1, 2, 3, and 4 represent progressively higher temperatures. Consider curve 1. It intersects at three points, A, A′, A″. The root at A″ has $c = \frac{1}{2}$. As was the case for Fig. 4.3, A and A″ are stable solutions; A′ is unstable. From symmetry, since A′ has $c = \frac{1}{2}$, we see that A and A″ correspond to the same value of the Gibbs free energy. If the temperature is raised slightly, the straight line becomes more horizontal, and we move to curve 2. The intersections near B and B′ are closer to each other than are B′ and B″. The solution at B″ then has the lower free energy since it is farther from the maximum and thus is the thermodynamically more stable. Thus if the system were at point *A* and were always in thermal equilibrium on increasing the temperature, c would take a jump to point B″. *The system*

[*Refs. on p. 59*]

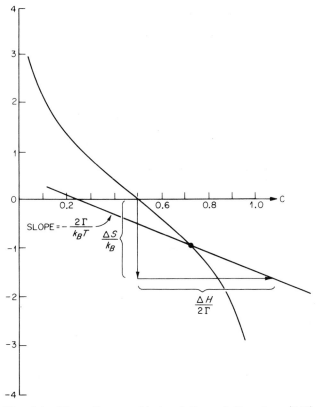

Fig. 4.4 Alternative graphical solution of Equation (4.29) showing location of a point which does not move with temperature (if Γ is independent of temperature).

would possess a discontinuous jump in c on raising or lowering temperature whenever the temperature corresponding to curve 1 were passed.

It is conceivable, however, that a system initally at A might not make the jump to A″ but rather remain *metastably* in the local free energy minimum. Such a situation could persist as the temperature is raised until curve 3 is reached. At point D the maximum and local minimum coincide. For any higher temperature, as with curve 4, there is only *one* intersection point and the system is forced to jump to point E″.

On lowering the temperature, such a metastable system starting at point E″ could pass progressively through D″, B″, and A″. When it reaches point F″, corresponding to curve 5, a further decrease now forces a jump to the left or low *c* side, since again there is only one intersection. Thus if the system can remain in metastable minima, *not only is there a discontinuous jump in c but there is hysteresis.*

[*Refs. on p. 59*]

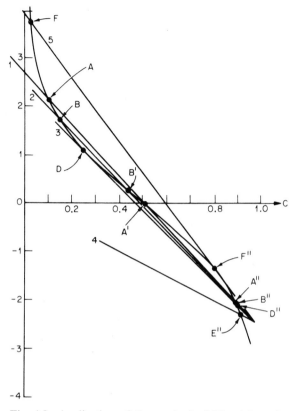

Fig. 4.5 Application of the method of Fig. 4.4 to the case of an attractive interaction (positive Γ) showing the possibility of discontinuous jumps in c as T varies, and of hysteresis. Lines 1, 2, 3, and 4 are at progressively higher temperatures. Line 5 is at a *lower* temperature than line 1.

We note that these phenomena will *only* occur for positive Γ. They thus can be used to show the existence of an attractive interaction. König and Madeja [3, 4] have observed low spin to high spin transitions in ferrous phenanthroline and ferrous bipyridyl compounds which have been something of a mystery because the site fraction of either species changes almost discontinuously with temperature. They have proposed the possibility of co-operative phenomena. We agree with their conjecture. We believe their experiments may arise from a case such as Fig. 4.5. We note that since they do not observe hysteresis their transitions all arise for the curve equivalent to curve 1 of Fig. 4.5. Clearly for this temperature

$$\Delta H = T\,\Delta S. \tag{4.30}$$

[*Refs. on p. 59*]

Moreover, for there to be a jump, the straight line must be steeper than the central region of the log curve

$$\Gamma \geq 2k_B T. \tag{4.31}$$

These facts alone enable us to conclude that for $Fe(phen)_2(NCS)_2$ $\Gamma/k_B \geq 358$ K, for $Fe(phen)_2(NCSe)_2$ $\Gamma/k_B \geq 464$ K, and for $Fe(dip)_2$ $(NCS)_2$ $\Gamma/k_B \geq 430$ K.

Our graphs so far have involved straight lines with negative slope corresponding to attraction. The contrast between attraction and repulsion can be shown on Fig. 4.6. The 'origins' P_1 and P_2 correspond to identical ΔS and ΔH's, but Γ's which are equal and opposite. Curves 1 and 2 correspond to one temperature, 3 and 4 to another. Thus we can compare the result of raising the temperature for two systems for which

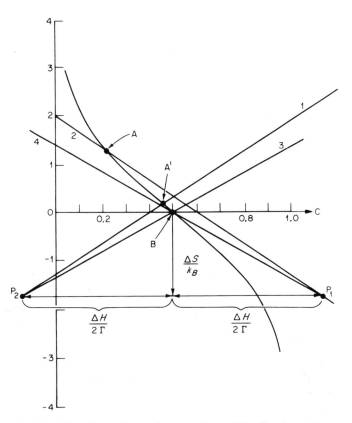

Fig. 4.6 The effect of changing the sign of Γ, all other things being equal. Lines 1 and 2 are at one temperature; lines 3 and 4 are at another (higher) one.

[*Refs. on p. 59*]

everything is equal except for a change in sign of Γ. For *attraction* we move from point A to point B but for *repulsion* the *same* temperature change moves us merely from A' to B, a much smaller shift in c. Thus we see graphically how the *repulsive* coupling produces a form of negative feedback which *inhibits* the changes in c, whereas *attraction* produces positive feedback which *enhances* changes in c.

So far, in discussing temperature change, our graphs have neglected any temperature variation in Γ. As a matter of fact constructions of Figs. 4.2 and 4.4 are completely valid even when a temperature-dependent Γ is used. The only point is that the 'temperature-independent point' of Fig. 4.4 would move with temperature. That a temperature variation of Γ is theoretically possible is evident from the definition of Γ as a contribution to the Gibbs free energy.

As a first approximation we treat the pressure and temperature variation of Γ by expanding in a power series about some temperature T_0 and pressure $p = 0$ keeping only the leading terms. We get

$$\Gamma(p, T) = \Gamma(0, T_0) + (T - T_0)\left(\frac{\partial \Gamma}{\partial T}\right)_{0, T_0} + p\left(\frac{\partial \Gamma}{\partial p}\right)_{0, T_0} \qquad (4.32)$$

which can be rewritten as

$$\Gamma(p, T) = \Gamma_a(p) + T\Gamma' \qquad (4.33)$$

defining the temperature-independent quantities Γ_a and Γ'. Of course we expect Equation (4.33) to be a good approximation within the same range of T near to T_0 for which Equation (4.32) is valid, even though T_0 has disappeared explicitly from Equation (4.33). Note, to this approximation, only Γ_a depends on pressure.

Using this form of Γ it is straightforward to show that the straight line representing the right hand side of Equation (4.29) still has a slope as a function of c given by

$$\text{slope} = -\frac{2\Gamma}{k_B T} = -\frac{2}{k_B}\left[\frac{\Gamma_a(p)}{T} + \Gamma'\right] \qquad (4.34)$$

and that at all temperatures for fixed pressure it passes through the point

$$\left[\frac{1}{2} + \frac{\Delta H}{2\Gamma_a}, -\frac{1}{k_B}\left(\Delta S + \frac{\Gamma'}{\Gamma_a}\Delta H\right)\right].$$

We shall denote this as the 'temperature-independent point'. We thus find that since ΔH and Γ_a are pressure-dependent one apparent effect of a

[*Refs. on p. 59*]

temperature-dependent Γ in terms of Fig. 4.4 is to make the entropy differences ΔS look to be pressure-dependent.

The expression for the temperature-independent point shows that its co-ordinates depend parametrically on pressure since both ΔH and Γ_a in general are functions of pressure. We neglect any pressure-dependence of Γ' or ΔS. Noting then that ΔH and Γ_a occur in just the ratio $\Delta H/\Gamma_a$ we see that as pressure is varied the temperature-independent point moves on a straight line of slope $-\Gamma'/2k_B$ passing through the point $(\frac{1}{2}, -\Delta S/k_B)$.

This situation is illustrated in Fig. 4.7. While it is difficult to obtain sufficiently accurate data to test these relationships, a temperature dependent Γ appears to be a necessary feature for a number of systems.

The general nature of these solutions should make them widely applicable. The graphical methods are particularly useful for understanding qualitatively co-operative effects in interacting systems as a function of temperature and pressure.

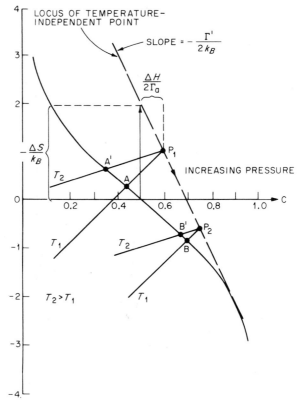

Fig. 4.7 Graphical method for the case that Γ varies linearly with T.

[*Refs. on p. 59*]

References

1. C. P. SLICHTER and H. G. DRICKAMER, *J. Chem. Phys.*, **56** 2142 (1972).
2. F. D. MURNAGHAN, *Finite Deformation of an Elastic Solid*, Dover Press (1967).
3. E. KÖNIG and K. J. MADEJA, *Inorg. Chem.*, **6** 48 (1967); *J. Amer. Chem. Soc.*, **90** 1146 (1968); *Spectra Chim. Acta*, **23A** 45 (1967).
4. E. KÖNIG, *Coord. Chem. Rev.*, **3** 471 (1968).

CHAPTER FIVE

Methods for Studying Electronic Structure at Very High Pressure

In this chapter we discuss the methods which have been extensively used to investigate electronic behavior in the pressure range to several hundred kilobars. We touch on the actual physical nature of the apparatus only to the degree necessary to give an idea as to the possibilities and limitations. The details necessary for construction of the apparatus appear in the literature [1-8]. We confine ourselves primarily to techniques used in this laboratory because we understand their possibilities and limitations. There exist more general reviews of high pressure technique [9-11].

The measurements used in this type of work to pressure in the range under discussion include optical absorption, electrical resistance, X-ray diffraction, and Mössbauer resonance. In general, the emphasis is on measuring intensive properties on very small samples, using the smallest high pressure cells possible.

Two not unrelated types of geometry are utilized: the Bridgman anvil (Fig. 5.1(a)) and the supported taper cell (Fig. 5.1(b)). The Bridgman anvil [12] makes use of the principle of 'massive support,' i.e., the fact that a material can be locally stressed well beyond the yield point if the member as a whole is below its yield point. The tapered piston cell involves Bridgman anvils with a compressible medium along the taper. As the pressure is applied the medium supports the center section and the piston with a pressure which decreases radially. It is as if the center section were a high pressure cell supported by a series of surrounding cells, each at a lower pressure. The pressure at the edge is sufficiently low and geometry sufficiently simple so that one can have small windows for the entrance and exit of light, X-rays, γ-rays, etc.

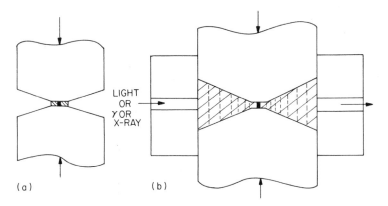

Fig. 5.1 (a) Bridgman anvils and (b) the supported taper apparatus.

The Bridgman anvil has the advantage of simple loading and operation and ready access for electrical leads and for radiation; however, it is considerably more limited in pressure range and is more susceptible to pressure gradients on the sample than is the tapered piston cell. The tapered piston cell has more uniform pressure on a small sample in the center of the flat [6] and is capable of a considerably higher pressure range; but it is more difficult to load, and somewhat more difficult to calibrate [8]. Either piece of apparatus is small in scale with all parts inexpensive to replace.

The pressure range in either apparatus depends on the material used for pistons and its treatment and on the pressure transmitting medium and support medium. The material used in the laboratory at the University of Illinois is sintered tungsten carbide with 3% cobalt binder. The material is work-hardened by stressing it well beyond the yield point and then regrinding. This considerably extends the pressure range, particularly for the supported taper apparatus. The pressure, especially in the supported taper apparatus, is limited by flow of the work-hardened carbide. The pressure range could be considerably extended by the use of sintered diamond pistons, or even sintered diamond tips on the anvils.

Typical diameters are 1.2 to 2.4 cm. The Bridgman anvil flats are 0.2 to 0.4 cm in diameter. The flats on the tapered piston cell are 0.1 to 0.2 cm in diameter. When the supporting material is pyrophyllite or pyrophyllite and LiH an 18° taper is used. For the optical cell where NaCl is the supporting material the taper is limited to 6° because of the greater tendency for radial flow of the medium. Center thicknesses range from 0.01 to 0.035 cm depending on the type of measurement involved. The material under study is typically confined at the center of the flat

[*Refs. on p. 70*]

in a cross section which is 5 to 10% of the flat area to minimize pressure gradients and resultant shear effects.

For electrical work, the support material is pyrophyllite, with the sample surrounded by a thin layer of AgCl to minimize shear. In the optical work both the pressure transmitting 'fluid' and the support medium are NaCl, while the X-ray and Mössbauer resonance work utilize a LiH-boron mixture, or, in some areas, pure LiH or pure boron.

Electrical work in Bridgman anvils is possible to a pressure of perhaps 150 kbar. The LiH-boron mixture allows X-ray and Mössbauer measurements to 180 to 200 kbar. In the supported taper cell electrical resistance work has been done to 500 kbar, X-ray diffraction to over 400 kbar and Mössbauer resonance to 300 kbar. The optical absorption work is limited to 150 kbar by flow of the NaCl medium. The X-ray and Mössbauer work is limited by dishing of the pistons which cuts off radiation. Calibration is based on the change of lattice parameters observed by X-ray diffraction [8, 13]. Densities obtained in shock wave experiments and corrected to room temperature are used. A number of different materials such as Al, NaCl, and MgO give consistent results. The optical cell is calibrated by observing optical effects associated with first-order phase transitions which are cross calibrated with the X-ray results.

It should be noted that to prepare material for chemical characterization, as discussed in Chapter 11, a supported taper cell is used with 0.32 cm flats and 0.07 cm height. This permits the use of a sample 0.16 cm in diameter and 0.07 cm high at pressures to 300 kbar [14].

It is often of some interest to mention the physical characteristics of the samples studied by the various techniques. In the electrical work, the metallic samples are normally strips approximately 0.08 to 0.1 cm long, 0.03 cm wide and 0.004 cm thick. Where possible, insulators and semiconductors are single crystal material shaped as nearly as possible to the above. On occasion pressure-fused samples are used, although they give more qualitative results. In the optical studies it is possible to use single crystals with careful loading. These are generally cut ~ 0.1 cm long and with thickness appropriate to the intensity of the absorption peak. More usually the material is powder diluted with NaCl and packed in a slot ~ 0.1 cm long by 0.03 to 0.04 cm in the direction of the light path. The dilution may be as low as 10 : 1 for weak absorption peaks and as high as 200 or 300 : 1 for intense peaks. The details of loading the optical and electrical cells are well illustrated in [5]. A better calibration for the electrical cell appears in [8].

[*Refs. on p. 70*]

In the X-ray and Mössbauer resonance work the sample was generally diluted with boron or boron and LiH, although other diluents did not affect the results. For the X-ray the dilution is 2 : 1 to 10 : 1; for the Mössbauer resonance the dilution is 6 : 1 to 30 : 1 depending on the amount of iron in the sample and the recoilless fraction. In these techniques special care is exercised to confine the sample to a very small fraction (5 to 6%) of the flat area, at its center, in order to minimize pressure gradients. There appears to be relatively little broadening of Mössbauer or X-ray peaks with pressure.

There are of course a number of high pressure apparatuses which have wide application but which we do not discuss here. These include the large scale gear like the General Electric belt [15] and the tetrahedral anvil equipment [16, 17], developed by Hall. These take relatively large volumes to 100 kbar or above at temperatures to 1000 to 1500 °C. McWhan and colleagues [18, 19] have developed equipment for making electrical measurements to 100 kbar down to liquid helium temperature. This apparatus was used for the work on transition metal oxides discussed in Chapter 7. Jayaraman and Maines [20] have perfected a technique for electrical measurements to 50 kbar using a hydrostatic medium. The possibilities of the 'collapsing sphere' technique invented by von Platen [21] and developed by Kawai [22] have not been fully explored as of now.

The temperature range for the tapered piston cell has so far been limited. The electrical resistance cell has been operated over the range 77 to 300 °K; the Mössbauer resonance gear over the range 80 to 425 K. The optical cell can in principle be operated in the range 77 to 425 K, but relatively few data have been taken far from room temperature.

It is useful to review the type of information concerning electronic behavior and electronic transitions which can be obtained from each of these techniques.

X-ray diffraction has been used extensively [13] to measure lattice parameters as a function of pressure for cubic and hexagonal crystals. It should be mentioned that Jamieson and Lawson [23] and McWhan [24, 25] have developed refined X-ray techniques using Bridgman anvils. Takahashi and Bassett [26] use a very elegant technique involving diamond anvils and a microfocus X-ray tube, which permits very precise X-ray film work to 300 kbar. Each pressure point requires 30 to 300 h, while in the supported taper cell the entire range can be covered in a few hours, but the improved precision of the diamond cell using film compensates for this, especially for more complex phases or for resolving

[*Refs. on p. 70*]

the structure of high pressure phases. Many of the materials involved in electronic transitions at high pressure are too complex for the available X-ray techniques. The major contribution to the work under discussion in this book is for the pressure calibration of the high pressure cells.

The electrical resistance cell is a two-lead system, which eliminates the possibility of obtaining resistivities. In metals one can detect the direction of the resistance change with pressure and the presence of maxima and similar anomalies. First-order phase transitions usually involve a discontinuity in resistance. Structural changes which involve atomic diffusion usually become markedly more sluggish as the temperature is lowered well below room temperature, so that true thermodynamic equilibrium is difficult to establish. Electronic transitions usually do not offer this problem and may even become sharper at low temperature. One can make a reasonable measurement of the temperature coefficient of resistance – this is helpful in establishing metallic character.

For insulators and semiconductors, the resistance may decrease by orders of magnitude with increasing pressure. The cell has a leakage resistance of about 10^8 Ω, so that even with very thin samples and short resistance paths it may not be possible to measure resistance at low pressure for good insulators. The resistance of such materials decreases exponentially with increasing temperature according to Equation (2.20) and one may make a reasonable measurement of the change in energy gap with pressure. An insulator-metal transition or semiconductor-metal transition is detectable because of the change in the temperature coefficient of resistance. These transitions are also frequently accomplished by a large discontinuous decrease in resistance. At the Research Institute for Swedish National Defense the resistance cell has been modified [27] to permit the introduction of up to six leads. This allows resistivity measurements, and opens up the possibility of Hall effect and related studies to very high pressure.

Optical absorption is probably the most generally useful tool available for very high pressure research on electronic behavior in insulators and semiconductors. Basically one measures the difference in energy between the ground state and an excited state subject to certain selection rules and other restrictions. These factors have been discussed earlier. The measurement of changes in this energy difference with pressure is essential to understanding the effect of pressure on electronic processes in general, and the nature of electronic transitions in particular. In the next chapter we review examples of a wide variety of such measurements,

[*Refs. on p. 70*]

including studies of the energy gap between the valence band and conduction band of semiconductors, various excitations localized on atoms, ions, or molecules, and the energy involved in charge transfer between entities in a solid.

The optical cell used in this work is capable of making absorption measurements with a minimum of shear on the sample to 150 kbar pressure over a wave length range from below 0.25 μ to well over 5 μ. The spectrometer slits must be open reasonably wide (usually 0.4 to 0.6 mm) so that fine structure is difficult to resolve. Most of the measurements to date have concerned only absorption edges or peak locations. With modern developments in light sources, detectors, and electronics, it is possible to establish changes in peak shape and in peak width with pressure. Absolute integrated intensities are very hard to measure. One can measure changes in the area under an absorption peak with pressure, but changes of cell geometry may mask area changes of 20 to 30% or more. Where there are large changes in integrated intensity with pressure rather accurate results can be obtained, especially if there is another peak in the spectrum which can be used as a fiducial marker. In addition to the optical absorption measurements, studies of fluorescence, phosphorescence, and Raman scattering [28-32] have been made in cells of this general design.

A very useful diamond cell for optical work has been developed by Wier *et al* [33]. It is especially valuable for work in the far infra-red, or for work with liquid pressure media in the 30 kbar range. For solid state work above about 40 to 50 kbar, it is difficult to avoid large pressure gradients on the sample in this cell.

Mössbauer resonance has been a very effective tool in identifying the electronic ground state of iron and in establishing in a semiquantitative way the relative amounts of two states where they exist together. Since the chemistry of iron is especially rich in electronic transitions Mössbauer spectroscopy is an important high pressure tool. It does not, of course, have as general an applicability to high pressure studies of electronic behavior as does optical absorption; however because its possibilities and limitations are perhaps less familiar to the chemical reader, and because Mössbauer measurements are discussed extensively in Chapters 8, 9, and 10, a rather lengthy exposition of the technique is necessary. There are several books [34-36] and many articles explaining the Mössbauer effect. Nevertheless, we shall outline the salient features with emphasis on those aspects important in detecting electronic transitions. In Mössbauer spectroscopy one measures the difference in energy between an excited

[*Refs. on p. 70*]

state and ground state of a nucleus (in our case ^{57}Fe) by measuring the energy of an emitted γ-ray. Because of a weak coupling between nuclear and electronic wave functions, one can use this measurement to infer information about the electronic ground state.

When a free radioactive atom decays giving off a γ-ray, the nucleus recoils to conserve momentum. In so doing, it carries off some of the energy of the transition so that the γ-ray is not in resonance with a second atom of the same type in the same chemical state. If the atom or ion is fixed in a crystal such that the lowest vibrational quantum (lowest phonon energy) of the lattice is large compared with the recoil energy, there will be a certain fraction of recoilless decays, and these will constitute a very sharp measure of the energy of the nuclear transition. This sharpness is essential, as one must be able to measure very small perturbations. The nuclear energy levels are slightly perturbed by electronic wave functions having non-zero amplitude at the nucleus (s electron wave functions). Since, in general, different chemical states or environments will expose the nucleus to differing amounts of s electron wave function, ^{57}Fe in, say, a stainless steel source will not be in resonance with ^{57}Fe in a ferric chloride absorber. If one moves the source with respect to the absorber (or vice versa) one can utilize the Doppler energy of motion to establish resonance. One then develops a scale so that the relative motion necessary to establish resonance measures the perturbation of the nuclear levels and thus the electron density at the nucleus. This is called the isomer shift or chemical shift. We use the former term and express all values relative to metallic bcc iron.

In order to discuss the use of this tool to study iron, it is necessary to digress briefly to outline its electronic structure. The free atom has filled atomic shells through the 3p orbitals. The outer structure is $3d^6$ $4s^2$. (Metallic iron may be nearer $3d^7$ $4s^1$.) Ferrous iron has the nominal outer structure $3d^6$ $4s^0$ and ferric iron $3d^5$ $4s^0$. The arrangement of the electrons in the partially filled 3d shell is an essential feature of iron chemistry. Fig. 5.2 shows possible arrangements for ferric and ferrous ions in octahedral symmetry. In the free atom, according to Hund's rule, the electrons are arranged to give maximum multiplicity (high spin). This arrangement also obtains for 'normal ionic' compounds. The high spin ferric state with five orbitals each containing one electron is a spherically symmetric ($^6A_1$) state. The high spin ferrous state is in an asymmetric ($^5T_2$) ground state. On the other hand, the low spin ferrous state is spherically symmetric ($^1A_1$), while the low spin ferric state ($^2T_2$) is

[*Refs. on p. 70*]

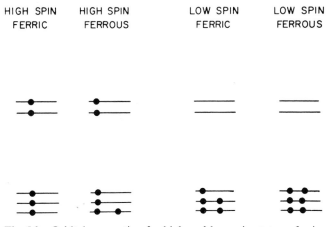

Fig. 5.2 Orbital occupation for high and low spin states – ferric and ferrous ions.

not. The symmetry of the ground state is an important feature for its identification, as is discussed below.

We are concerned with changes in the isomer shift accompanying the chemical bonding. The 1s and 2s electrons provide an electron density at the nucleus which does not depend on the chemical state. The 3s electrons do not interact directly with the surroundings, but their radial maximum occurs at about the same point as that of the 3d orbitals. Electrons in the 3d orbitals will thus shield the 3s electrons. This provides a first order difference between ferric and ferrous iron. The amount of shielding will also depend on the radial extent of the 3d orbitals. They may be spread out by covalent interaction in compounds or by the band structure of the metal.

The isomer shift of the metal, then, depends on the filling and radial extent of the 3d orbitals and also of the 4s orbital. In compounds of iron the decisive features are the oxidation state and the spreading of the 3d orbitals due to covalency. Particularly in ferric iron there is also the possibility of some occupation of the 4s orbitals. In addition, it is possible for ligand orbitals to overlap and shield the 3s electrons. These features are discussed in more detail in Chapter 6.

There is a second feature of the Mössbauer spectrum which helps in our identification of the chemical state. It is possible for an electric field gradient to interact with the nuclear quadrupole moment to split the excited state of spin 3/2 and give two resonances, as illustrated in Fig. 5.3. An electric field gradient may be caused by either an aspherical occupation of the 3d shell or by a non-cubic arrangement of the ligands. Since

[*Refs. on p. 70*]

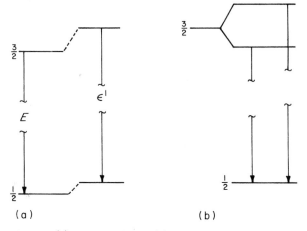

Fig. 5.3 (a) Isomer shift and (b) quadrupole splitting.

quadrupolar forces are short range, and the 3d orbitals are much closer to the nucleus than the ligand orbitals, the former effect, when present, is dominant.

In Fig. 5.4 we show typical spectra for the three states which we shall encounter most frequently in this work. The high spin ferrous state involves a large positive isomer shift (1.0 to 1.4 mm s^{-1}), corresponding to a low electron density at the iron nucleus because of the shielding of the 3s electrons by 3d electrons. The large quadrupole splitting (2.0 to 3.0 mm s^{-1}) corresponds to the aspherical occupation of the 3d shell. High spin ferric iron exhibits a significantly lower isomer shift (0.3 to 0.5 mm s^{-1}) primarily due to the smaller shielding. There is usually enough deviation from cubic symmetry in the ligand arrangement to give a small (0.5 to 0.7 mm s^{-1}) quadrupole splitting. The low spin ferrous ion exhibits a very low isomer shift (0.0 to 0.1 mm s^{-1}) because the 3d orbitals are strongly delocalized. This is discussed in some detail in a later chapter. The quadrupole splitting is small because of the spherically symmetric occupation of the 3d shell.

In the high pressure Mössbauer resonance experiments the sample is made from iron enriched to 70 to 95% in ^{57}Fe, and the compounds are diluted (5 : 1 to 10 : 1) with boron and pressed into a hole 0.03 cm in diameter in a pellet of boron plus LiH 0.2 cm in diameter (other diluents such as Al_2O_3 or graphite do not seem to affect the results). The problems in making quantitative measurements and in their interpretation fall into three categories: those connected with Mössbauer resonance, those connected with non-hydrostaticity and shear, and those

[*Refs. on p. 70*]

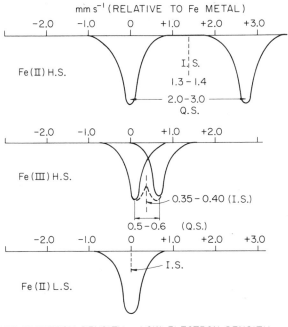

Fig. 5.4 Typical Mössbauer spectra.

associated with metastability, which may be inherent in solid state reactions.

The relative amounts of the various states are established from the area measured by fitting the data with Lorentzian or pseudo-Lorentzian peaks. It is necessary to assume equal f number at the two sites. This is always a possible source of error. Self absorption in the sample can change the apparent relative amounts of two states. This may be a source of significant error in the earlier work where the degree of dilution was not closely controlled. If high dilutions are used and care is taken to run a series of similar compounds at the same dilution, this error can be largely eliminated. A third possible inaccuracy arises from fitting asymmetric peaks, as in the hemiporphyrins. Under the best conditions the conversions can be reproduced to $\sim \pm 2\%$.

The pressure is clearly not completely hydrostatic. Under these conditions there is a question of the degree to which conversions are due to shear rather than pressure. There are a number of reasons for believing that pressure *per se* is the controlling factor in the great majority of cases:

[*Refs. on p. 70*]

(1) Isobaric runs show a large effect of temperature on conversion at constant pressure. With no change in pressure, change in shear should be minimized.

(2) Studies involving a series of related compounds with apparently similar shear properties sometimes gave very different conversions, and conversions which correlated well with other electronic differences, as is discussed in later chapters.

(3) For some compounds with distinctive shear products it is possible to make pressure runs where little or no shear product is produced at any pressure.

(4) The change in integrated intensity (area) under optical charge transfer peaks with pressure mirrors the conversion obtained from Mössbauer studies. Shear is minimized in the optical apparatus.

(5) The conversions were generally in some sense reversible. When the applied pressure is removed there is still considerable strain in the pellet. However, for some systems (hemin, hematin) the spectrum on releasing all the applied pressure was substantially identical with the atmospheric. For most systems the spectrum has returned 60 to 80% of the way to the original. It is possible to cut the center from the pellet and to relieve the strain by chopping it with a sharp blade, although with such a small amount of material the efficiency of the operation is low. This chopped material exhibits an 80 to 100% return to the original spectrum for most materials.

Except possibly for one or two systems, it is probable that the lack of immediate reversibility is primarily due to stored strain in the crystals because of modified local geometry, especially at high conversions. This raises the question of establishing true equilibrium in solid state systems where electron transfer or other chemical processes involving local deformation occur. The conversions are not time-dependent in the sense that readouts over a period of time at the same pressure give the same conversion, even over periods of several days. However, if two states of only moderately different energy are separated by a reasonably high potential barrier, the system could stay metastably in the higher energy state for an indefinite period.

References

1. R. A. FITCH, T. E. SLYKHOUSE and H. G. DRICKAMER, *J. Opt. Soc. Am.*, **47**, 1015 (1957).
2. H. G. DRICKAMER, *Rev. Sci. Insr.*, **32**, 212 (1961).
3. A. S. BALCHAN and H. G. DRICKAMER, *Rev. Sci. Instr.*, **32** 308 (1961).

4. E. A. PEREZ-ALBUERNE, K. F. FORSGREN and H. G. DRICKAMER, *Rev. Sci. Instr.*, **35** 29 (1964).
5. H. G. DRICKAMER and A. S. BALCHAN in *Modern Very High Pressure Techniques*, p. 25, edited by R. H. Wentorf, Jr., Butterworths, London (1962).
6. K. F. FORSGREN and H. G. DRICKAMER, *Rev. Sci. Instr.*, **36** 1709 (1965).
7. P. DE BRUNNER, R. W. VAUGHAN, A. R. CHAMPION, J. COHEN, J. MOYZIS and H. G. DRICKAMER, *Rev. Sci. Instr.*, **37** 1310 (1966).
8. H. G. DRICKAMER, *Rev. Sci. Instr.*, **41** 1667 (1970).
9. C. A. SWENSON in *Solid State Physics*, Vol. 11, edited by F. Seitz and D. Turnbull, Academic Press, New York (1960).
10. R. H. WENTORF, JR. (ed.), *Modern Very High Pressure Techniques*, Butterworths, London (1962).
11. C. C. BRADLEY, *High Pressure Methods in Solid State Research*, Plenum Press, New York (1969).
12. P. W. BRIDGMAN, *Proc. Amer. Acad. Arts and Sci.*, **81** 165 (1952).
13. H. G. DRICKAMER, R. L. CLENDENON, R. W. LYNCH and E. A. PEREZ-ALBUERNE in *Solid State Physics*, Vol. 19, edited by F. Seitz and D. Turnbull Academic Press, New York (1966).
14. V. C. BASTRON and H. G. DRICKAMER, *J. Solid State Chemistry*, **3** 550 (1971).
15. H. T. HALL, *Rev. Sci. Instr.*, **31** 125 (1960).
16. H. T. HALL, *Rev. Sci. Instr.*, **29** 267 (1958).
17. E. C. LLOYD, U. O. HUTTON and D. P. JOHNSON, *J. Res. Nat. Bur. Standards*, **63c** 59 (1959).
18. D. N. LYON, D. B. MCWHAN and A. L. STEVENS, *Rev. Sci. Instr.*, **38** 1234 (1967).
19. D. B. MCWHAN, T. M. RICE and P. H. SCHMIDT, *Phys. Rev.*, **177** 1063 (1969).
20. A. JAYARAMAN and R. G. MAINES in *Accurate Characterization of the High Pressure Environment*, edited by E. C. Lloyd (NBS Pub. 326 Washington, D.C.)
21. B. VON PLATEN in *Modern Very High Pressure Techniques*, p. 118. edited by R. H. Wentorf, Jr. Butterworths, London (1962).
22. N. KAWAI and S. ENDO, *Rev. Sci. Instr.*, **41** 1178 (1970).
23. J. C. JAMIESON and A. W. LAWSON, *J. Appl. Phys.*, **33** 775 (1962).
24. D. B. MCWHAN, *J. Appl. Phys.*, **38** 347 (1967).
25. D. B. MCWHAN, *Trans. Amer. Cryst. Assoc.*, **5** 39 (1969).
26. T. TAKAHASHI and W. A. BASSETT, *Science*, **145** 483 (1964).
27. G. LEMAN and A. FRÖMAN, (personal communication).
28. H. OFFEN, *Rev. Sci. Instr.*, **39** 1961 (1968).
29. H. OFFEN in *Organic Molecular Photophysics*, edited by J. B. Birks, Wiley-Interscience, New York (in press).
30. D. GREGG and H. G. DRICKAMER, *J. Appl. Phys.*, **31** 494 (1960); *J. Chem. Phys.*, **35** 1780 (1961).
31. M. NICOL, *J. Opt. Soc. Am.*, **55** 1176 (1965).
32. M. NICOL and J. SOMEKH, *J. Opt. Soc. Am.*, **58** 233 (1968).
33. C. E. WIER, A. VAN VALKENBURG and E. LIPPINCOTT, in *Modern Very High Pressure Techniques*, p. 51, edited by R. H. Wentorf, Jr., Butterworths, London (1962).
34. H. FRAUENFELDER, *The Mössbauer Effect*, W. A. Benjamin, New York (1962).
35. G. WERTHEIM, *The Mössbauer Effect – Principles and Applications*, Academic Press, New York (1964).
36. V. I. GOLDANSKII and R. H. HERBER, *Chemical Applications of Mössbauer Spectroscopy*, Academic Press, New York (1968).

CHAPTER SIX

Shifts of Energy Levels with Pressure

As we have indicated in the introduction, a common effect of pressure is to shift the energy of one type of orbital with respect to another. Usually one measures the shift by optical absorption although other techniques, such as Mössbauer resonance, may give useful information. In this chapter we describe a variety of such shifts: relatively localized transitions such as those involving the d electrons of transition metals; excitations involving color centers in alkali halide crystals, or π electrons in aromatic hydrocarbon crystals; charge transfer between donors and acceptors in molecular complexes or between ligand and metal in transition metal complexes; and finally, excitations from the valence band to the conduction band in insulators and semiconductors. In the following chapters we discuss orbital energy shifts which lead specifically to new ground states in a variety of systems.

6.1 d-d transitions

In Chapter 2 we discussed the characterization of the field imposed by the ligands on a partially filled 3d shell in terms of the ligand field strength Δ and the Racah parameters B and C. We shall restrict ourselves in this section primarily to systems of essentially octahedral symmetry, and to high spin ions. In Fig. 6.1 the shifts of two peaks of Ni^{2+} in MgO [1] are shown as a function of pressure. The first measures Δ directly, while the second depends on both Δ and B. Δ increases with pressure, in qualitative agreement with theory. Fig. 6.2 exhibits the change of Δ with pressure for four ions in MgO. It is clear that the fractional change is not identical for the four ions.

[Refs. on p. 108]

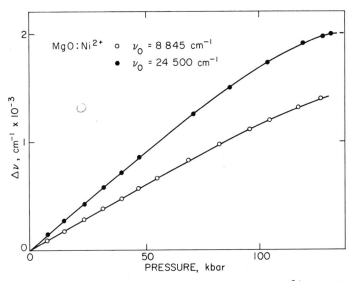

Fig. 6.1   Shifts of ligand field peaks versus pressure – Ni^{2+} in MgO.

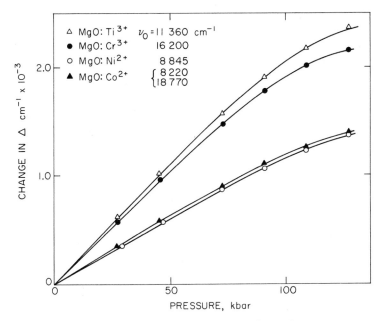

Fig. 6.2   Change of Δ with pressure – four ions in MgO.

[*Refs. on p. 108*]

In Chapter 2 we pointed out that a simple point charge theory would predict that the field at the ion would vary as R^{-5} where R is the ion-ligand distance. In Fig. 6.3 the fractional change in Δ is plotted against pressure, together with $(a_0/a)^5$ where the lattice parameter is established from X-ray measurements [2]. There is variation from ion to ion but, in general, the change in Δ is somewhat larger than predicted. On the other hand, similar data for ions in Al_2O_3 exhibited in Fig. 6.4 show remarkable agreement [3, 4]. It is not clear, for foreign ions in a crystal lattice, whether or not the local compressibility near the foreign ion varies significantly from the bulk compressibility. Fig. 6.5 compares the prediction from p-V data [5] with the measured values of Δ for NiO. The changes in the latter are slightly larger than predicted. Here there is no local relaxation effect, but NiO is very far from a purely ionic crystal. The point charge model is, of course, inadequate to calculate Δ at 1 atm. Even for semiquantitative calculation in ionic crystals like the fluorides molecular orbital theory with configuration interaction is required [6]. Under these circumstances, it is perhaps surprising that the theory gives even the right magnitude of the pressure change. It may be coincidental, or it may be that, for these systems at least, the effects of pressure on the electrostatic component of Δ are larger than pressure effects on other aspects of the ligand field.

The Racah parameters also depend on pressure. In Fig. 6.6 B is plotted against pressure for Cr^{3+} in Al_2O_3 [7] (ruby). There is a significant decrease. For Mn^{2+} ions there exist transitions which depend on B and

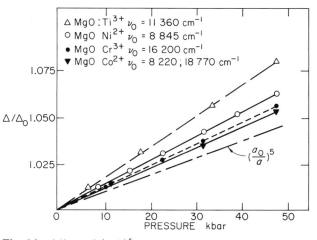

Fig. 6.3 Δ/Δ_0 and $(a_0/a)^5$ versus pressure – various ions in MgO.

[*Refs. on p. 108*]

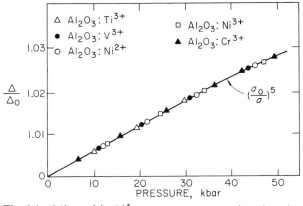

Fig. 6.4 Δ/Δ_0 and $(a_0/a)^5$ versus pressure – various ions in Al_2O_3.

C only; these peaks permit an accurate determination of these parameters. In Fig. 6.7 the change of B and C with pressure is plotted for $MnCl_2$ and $MnBr_2$ [8] which are tetrahedral complexes. Both parameters decrease with pressure – the fractional change of B is somewhat larger than that of C. The decrease of the Racah parameters is consistent with the

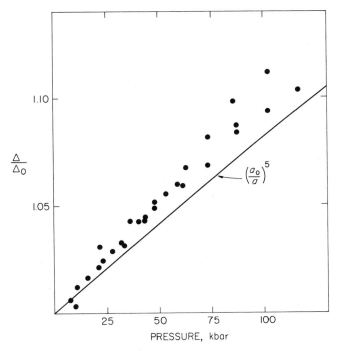

Fig. 6.5 Δ/Δ_0 and $(a_0/a)^5$ versus pressure – NiO.

[*Refs. on p. 108*]

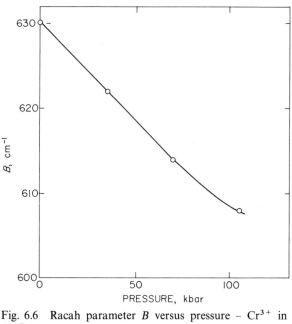

Fig. 6.6 Racah parameter *B* versus pressure – Cr^{3+} in Al_2O_3.

idea that the 3d orbitals tend to spread out with pressure (central field covalency). This increases the interelectronic distance and decreases the repulsion. This effect has implications for the change of isomer shift with pressure discussed below, and also for the energy of the 3d orbitals *vis-à-vis* the ligand non-bonding orbitals, and thus for the reduction of ferric iron discussed in Chapter 9.

It is of interest to note that for the very covalent system $K_3Co(CN)_6$ there is evidence that the Racah parameter *B* increases with pressure [9]. Possibly some relocalization of electrons takes place. This may be related to the change in spin state of ferrocyanides discussed in Chapter 8.

In all of the theory we have discussed it is assumed that the transition metal ion is perturbed by its nearest neighbor ligands, which is certainly a reasonable first-order assumption. However, in Fig. 6.8 we plot the shift of the peak measuring Δ for Ni^{2+} in $Ni(NH_3)_6Cl_2$. Here Ni^{2+} is surrounded by six (NH_3) molecules in octahedral symmetry. Near 60 to 65 kbar there is a phase transition which presumably rearranges the nickel ammonium molecular ions with respect to the Cl^- ions. There is a small but measurable discontinuity in the peak shift [10] and, therefore, in the field. Similar discontinuities have been observed in other systems at phase

[*Refs. on p. 108*]

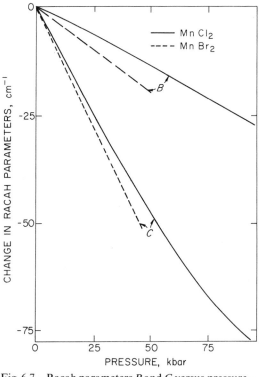

Fig. 6.7   Racah parameters B and C versus pressure –
$MnCl_2$ and $MnBr_2$.

transitions. The rearrangement of the ions evidently changes the degree of compression of the nickel ammonium ions, and therefore it changes the field at the Ni^{2+}.

6.2   Isomer shift

As indicated in Chapter 5, in a Mössbauer spectroscopy experiment the resonance velocity of the source with respect to the absorber can be related to the s electron density at the nucleus for the Mössbauer isotope. For ^{57}Fe one can then relate the isomer shift to the degree of occupation and radial extent of the 3d orbitals (oxidation state and degree of covalency). Indeed, as pointed out by Erickson [11], the isomer shift depends on a variety of effects. In the first place, it is affected by the 3d orbital expansion due to a reduction of the effective nuclear charge associated with the overlap of the metal electron cloud with the negative ligand charge; and in the second place, by 4s orbital augmentation.

[*Refs. on p. 108*]

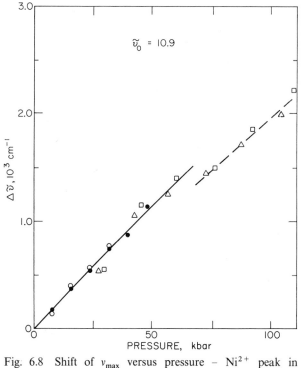

Fig. 6.8 Shift of v_{max} versus pressure – Ni^{2+} peak in $Ni(NH_3)_6Cl_2$.

These two effects constitute central field covalency. In the third place, metal d_π backbonding to vacant π^* orbitals can be an important source of 3d electron delocalization. This is an important contribution to reducing the isomer shift (and to increasing the ligand field) in many covalent metal-organic compounds. Finally, there is the possibility that ligand electrons will overlap the metal 3s orbitals and shield them from the nucleus. These latter two effects constitute symmetry-restricted covalency.

For relatively ionic compounds, one might anticipate that pressure would primarily act to increase the delocalization obtained through the first of these effects. It is indeed this delocalization which is the major cause of the decrease in the Racah parameters with pressure discussed in the previous section. One would then expect that, for these ionic compounds, one would observe a decrease in isomer shift (an increase in electron density at the nucleus) with pressure. As can be seen in Fig. 6.9, just such an effect is observed for ionic ferrous and ferric compounds. Changes in 4s occupation may contribute to this effect, but for these materials it is probably not a major factor.

[*Refs. on p. 108*]

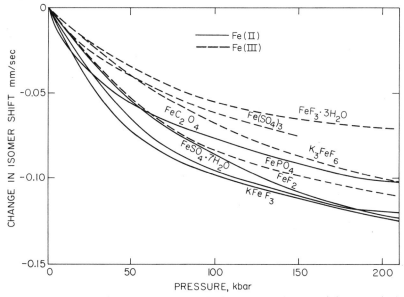

Fig. 6.9   Isomer shift versus pressure – ionic compounds of Fe (II) and Fe(III).

For studying both changes in the characteristics of the 3d electrons and changes in ground state (electronic transitions), it is desirable to study in a systematic way a series of related compounds with different degrees of covalency. For this purpose, we have studied the properties of the ferric complexes of the twelve β-diketones [12] shown in Fig. 6.10. By changing the nature and position of the side groups on the quasi-aromatic ring, one can vary the electronic properties of the oxygen and, therefore, change the nature of the metal-ligand interaction. While the molecular symmetry is D_{3d} or lower, the local symmetry at the iron is essentially octahedral. In Chapter 9 we discuss electronic transitions in these compounds and relate the degree of reduction to the electronic properties of the ligand. Here we discuss only the effect of pressure on the isomer shift of the ferric ion, and of the ferrous ion formed from it under pressure.

It is convenient to group the derivatives studied into smaller classes (A, B, and C) so that the pressure behavior may be discussed. These classifications are based on the isomer shift behavior of both oxidation states as a function of pressure above 40 to 50 kbar. They are not rigid, e.g., at low pressures PACA(10) is more like a Class A compound and at high pressures FTFA(6) approaches Class C behavior. Similarities in behavior of the Fe(II) isomer shifts will be considered first, followed by an examination of the consistencies in the Fe(III) isomer shifts.

[*Refs. on p. 108*]

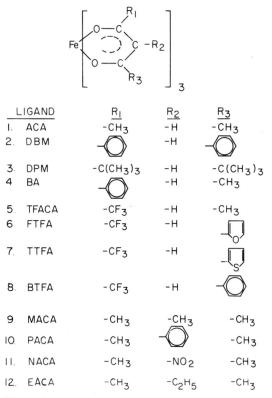

LIGAND	R_1	R_2	R_3
1. ACA	$-CH_3$	$-H$	$-CH_3$
2. DBM	⟨phenyl⟩	$-H$	⟨phenyl⟩
3. DPM	$-C(CH_3)_3$	$-H$	$-C(CH_3)_3$
4. BA	⟨phenyl⟩	$-H$	$-CH_3$
5. TFACA	$-CF_3$	$-H$	$-CH_3$
6. FTFA	$-CF_3$	$-H$	⟨furyl⟩
7. TTFA	$-CF_3$	$-H$	⟨thienyl⟩
8. BTFA	$-CF_3$	$-H$	⟨phenyl⟩
9. MACA	$-CH_3$	$-CH_3$	$-CH_3$
10. PACA	$-CH_3$	⟨phenyl⟩	$-CH_3$
11. NACA	$-CH_3$	$-NO_2$	$-CH_3$
12. EACA	$-CH_3$	$-C_2H_5$	$-CH_3$

Fig. 6.10   Structure of substituted β-diketones.

Some interesting observations may be made if the smoothed data for all the derivatives are plotted on a single figure, as shown in Fig. 6.11. The six derivatives drawn with solid curves, ACA(1), BA(4), TFACA(5), MACA(9), NACA(11), and EACA(12) will be referred to as Class A. These are closely grouped at low pressures with Fe(II) isomer shifts in the range 1.26 to 1.32 mm s^{-1}, typical of systems with little or no metal-to-ligand backdonation.

On the basis of Fe(II) isomer shift data, the members of the second class are FTFA(6), TTFA(7), BTFA(8), and PACA(10). The Class B derivatives have distinctly smaller low pressure Fe(II) isomer shifts (1.06 to 1.14 mm s^{-1}) than the members of Class A. Such values indicate a considerably larger backdonation. However, the decrease of Fe(II) isomer shifts with pressure is not nearly as large as in Class A; the final values lie between 0.99 and 1.06 mm s^{-1}. This is due to a reduction in backdonation as explained in the following section on Fe(III) isomer shifts. Fe(III) isomer shift data for FTFA(6) and PACA(10) are somewhat
[*Refs. on p. 108*]

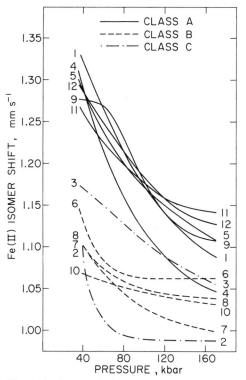

Fig. 6.11   Isomer shift versus pressure – Fe(II) in twelve β–diketones.

atypical for this group. The former exhibits properties similar to Class C at high pressure and the latter to Class A at low pressure.

The grouping of DBM(2) and DPM(3) in Class C can be rationalized better on the basis of the Fe(III) isomer shift than on the values for Fe(II). Although both DBM(2) and DPM(3) show lower Fe(II) isomer shifts than members of Class A, their smoothed values completely bracket those of Class B. The very low Fe(II) isomer shift of DBM(2) indicates rather extensive backdonation to the aromatic terminal substituents. Certainly the tertiary butyl groups in DPM(3) would not allow for as much delocalization.

The pressure behavior of Fe(III) isomer shifts can now be considered in terms of consistencies with and deviations from the classification which has been presented. The smoothed values for all of the derivatives are arranged by classes in Figs. 6.12, 6.13 and 6.14. As will be discussed in Chapter 9, the Fe(III) isomer shift at 1 atm is a good measure of the electron donating properties of the ligand. Thus it is reasonable that for Class A the electron donating methyl (MACA(9)) and ethyl

[*Refs. on p. 108*]

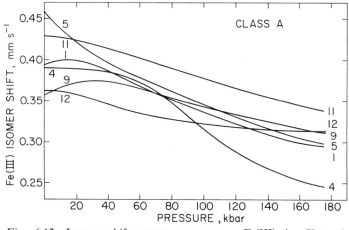

Fig. 6.12 Isomer shift versus pressure – Fe(III) in Class A
β-diketones.

(EACA(12)) derivatives should have low isomer shifts and that the electron withdrawing trifluoromethyl (TFACA(5)) and nitro (NACA(11)) derivatives have high isomer shifts, with ACA(1) and BA(4) intermediate. Because of the fairly broad range of properties, the atmospheric isomer shifts cover a rather large range, from 0.36 to 0.46 mm s^{-1}. The typical behavior of the ferric isomer shift, as discussed above, is to decrease with pressure because of an expansion of the d orbitals and subsequent reduction of shielding of the 3s orbital. The isomer shifts for the Class A compounds do show net decreases over the pressure range, but in the low pressure region, except for TFACA(5), they exhibit only very small decreases or even increases and subsequent maxima at moderate

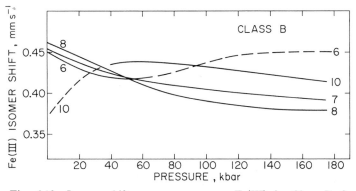

Fig. 6.13 Isomer shift versus pressure – Fe(III) in Class B β-diketones.

[*Refs. on p. 108*]

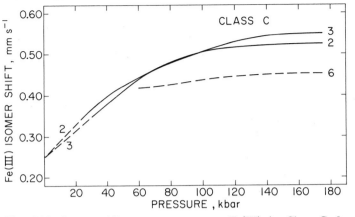

Fig. 6.14 Isomer shift versus pressure – Fe(III) in Class C β-diketones.

pressures. This behavior may be explained in terms of competition among the various contributions to the isomer shift. It is presumed that all of the derivatives have some 4s occupation in the ferric state, the extent varying from compound to compound. The inductive effects exhibited by the various substituents through the σ bonding system should be important in this respect, with good σ donors such as the alkyl groups increasing the 4s augmentation, and poor σ donors such as the trifluoromethyl group decreasing it. With pressure, the 4s orbital, which is σ antibonding, will increase in energy with respect to the metal 3d and ligand π and π* orbitals due to increased overlap with ligand σ orbitals. This will cause a decrease of 4s occupation and an increase in isomer shift which, of course, is counter to the result expected from a delocalization of the metal d electrons. Thus a competitive balance will be maintained for each compound and the possibility of maxima in the isomer shift curves is apparent.

In contrast to the other derivatives of Class A, TFACA(5), a strong σ acceptor with the highest initial isomer shift, shows a very large initial decrease in IS. This may be attributed to a very low initial 4s occupation so that the spreading of the d orbitals is the predominant pressure effect. The initial differences in behavior among the members of Class A are reduced above 100 kbar as the smoothed curves for four of the derivatives (ACA(1), TFACA(5), MACA(9), and EACA(12)) form a narrow band of about 0.02 mm s^{-1} width with NACA(11) lying about 0.03 mm s^{-1} above and BA(4) from 0.03 to 0.05 mm s^{-1} below.

The smoothed curves for Class B are shown in Fig. 6.13. The atmospheric and low pressure behavior of the three fluorinated derivatives

[*Refs. on p. 108*]

FTFA(6), TTFA(7), and BTFA(8) are very similar, with the fairly high values (0.45 to 0.46 mm s^{-1}) predominantly due to the presence of the electronegative trifluoromethyl group. There is no tendency toward maxima in the low pressure region for these three derivatives. This is consistent with the assumption that the trifluoromethyl group reduces the initial 4s augmentation to very small values. However, in contrast to the large TFACA(5) isomer shift decrease of 0.16 mm s^{-1} over the whole pressure range, TTFA(7) and BTFA(8) show much smaller decreases of 0.07 and 0.08 mm s^{-1}, respectively. FTFA(6) shows an increase at higher pressures. The similar behavior of TTFA(7) and BTFA(8) is reasonable on the basis of the structure of the second terminal substituent, where the sulfur in the heterocyclic thiophene ring of TTFA(7) has the same Pauling electronegativity as carbon in the phenyl ring of BTFA(8). The fourth member of the class, PACA(10), also exhibits only a rather small decrease in isomer shift above 50 kbar.

It has been noted already that, on the basis of Fe(II) isomer shifts, these derivatives show extensive backbonding in the ferrous state. Backbonding would be expected for the ferric state also but, in general, this state has less than the ferrous state. The relatively small decreases in isomer shift may then be associated with a decrease in backbonding with pressure such as that observed in phenanthroline complexes and some ferrocyanides, as discussed in Chapter 8.

PACA(10) shows typical Class B behavior above 50 kbar, but exhibits an unusually large maximum in the isomer shift at low pressure. No doubt this is in part due to a decrease in 4s augmentation with pressure, as occurs for most Class A complexes. In addition, there may be exceptionally large changes in the ligand σ-metal 3s overlap because of changing steric effects with this unusually large ligand.

The Fe(III) isomer shifts for DBM(2) and DPM(3), are shown in Fig. 6.14. Low pressure data are difficult to obtain because of the very low per cent effect and the non-Lorentzian broadening mentioned in Chapter 9. (This broadening diminishes with increasing pressure and becomes insignificant above 70 kbar). However, it appears that both isomer shifts are initially rather low, of the order of 0.25 mm s^{-1}. The low values probably arise from different effects with extensive d$_\pi$ delocalization to the two aromatic rings predominating in DBM(2) and appreciable σ donation from the tertiary butyl groups in DPM(3). The important feature is that both derivatives show quite large increases in Fe(III) isomer shift so that at the highest pressure the values range between 0.52 and 0.55 mm s^{-1}. As already mentioned, FTFA(6) also

[*Refs. on p. 108*]

shows an increase in isomer shift in the high pressure region, qualitatively similar to DBM(2) and DPM(3). The increase in isomer shift may be explained in part by the reduction of backbonding and 4s occupation with pressure for DBM(2) and DPM(3), respectively. In addition, there are probably contributions due to 3s shielding arising from overlap with occupied ligand orbitals.

This study illustrates the way that the isomer shift can be used to understand the relative displacement of the valence electrons with changing pressure and nature of the ligand. These displacements are important in understanding a variety of electronic transitions.

6.3   Color centers

Still another form of relatively localized excitation is the 'color center' in alkali halides. A wide variety of such centers may be introduced by X-irradiation, excess alkali metal ion, or excess halide ion. We shall confine our discussion here to the F center, which consists of a halide ion vacancy with an electron trapped in it. There are various reviews which discuss these and other types of centers [13, 14]. Pressure effects on a variety of color centers and other alkali halide impurities are discussed in some detail elsewhere [3]. The optical excitation is to a bound excited state of p, or possibly of mixed s-p character. The usual simple treatment approximates the center as a 'particle in a box' so that the energy is proportional to $1/A_0^2$, where A_0 is the local lattice parameter. Using peak energies and lattice parameters at 1 atm for a number of alkali halides, Mollwo [15] showed that this relationship is roughly correct. Ivey's [16] analysis of the data gives

$$EA_0^{1.84} = 1.76 \times 10^{-19} \text{ eV m}^2 \qquad (6.1)$$

where E is in electron volts and A_0 is in centimeters. Jacobs [17] showed that Ivey's equation is an approximation to a more general relationship:

$$\frac{\partial \ln E}{\partial \ln A_0} = \frac{3\gamma bT}{E\theta} + n_Q f_F \qquad (6.2)$$

where γ is the Gruneisen constant, θ is the Debye temperature, E is the energy of the F center, T is the absolute temperature, n_Q is the change in F center energy with local compressibility, f_F is the ratio of local to bulk compressibility, and b is an experimentally determined coefficient.

[*Refs. on p. 108*]

Jacobs showed that, at low pressures at least, the second term dominates. From elasticity theory he obtained an approximate relationship for f_F:

$$f_F = 1 + \frac{3}{4}\frac{B}{\mu} \tag{6.3}$$

where B is the bulk modulus and μ is the shear modulus.

Fig. 6.15 shows the fractional change in F center frequency with density for NaCl, NaBr and KCl [18, 19] using Bridgman's [20] density data. The line marked A_0^{-2} shows the particle in the box approximation. At low pressures $f_F = 2.1$ for NaCl and 2.3 for NaBr, which are close to Jacob's values. At high pressure f_F approaches 1.0 to 1.5. It is intuitively reasonable that the local compressibility should approach the bulk compressibility at high pressure as with increasing compression one would expect increased overlap among electron clouds of nearest neighbor ions.

One can show that Equation (6.2) predicts this qualitatively. The bulk modulus of NaCl increases by 60 to 70% from 10 to 50 kbar. In the same pressure range Bridgman's [21] results indicate that the 'resistance to shear' increases by about a factor of 3.3. From these values one would estimate an f_F of 1.6 at 50 kbar. For KCl the change in bulk modulus is 2.7 and the increase in resistance to shear is 5.2, giving f_F (50 kbar) $= 1.6$. These are qualitatively similar to the experimental values.

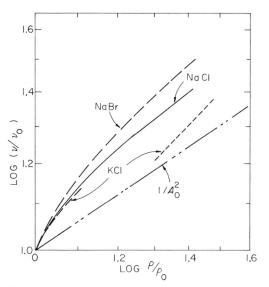

Fig. 6.15   Log v/v_0 versus log ρ/ρ_0 – F band in alkali halides.

[*Refs. on p. 108*]

6.4 $\pi \rightarrow \pi^*$ Transitions

As discussed earlier, π orbitals may exhibit considerably less overlap at atmospheric pressure than σ orbitals, and thus may be more pressure-sensitive. In this section we discuss the change in energy with pressure of the optical absorption maxima corresponding to π-π^* transitions (usually the lowest energy transition) in systems containing conjugated π bonds. We shall see that these energies are, indeed, very sensitive to pressure. Offen [22] has recently reviewed in detail pressure effects on absorption and emission of radiation from π orbitals at moderate pressures. We shall see that for a variety of cases the shift in energy is sufficient to give thermal occupation of the π^* state with pressure, sometimes with interesting chemical consequences. We discuss first the low energy transitions in the linear polyacenes anthracene, tetracene, and pentacene, and in azulene [23, 24]. (Azulene is an isomer of naphthalene with one five-membered and one seven-membered ring.)

In Fig. 6.16 is plotted the shift with pressure of the transition from the A_1 ground state to the first excited state of anthracene, labelled ^1L_a by Klevens and Platt [25]. The peaks represent different vibrational states. Tetracene and pentacene exhibit similar behavior. In Fig. 6.17 is plotted the shift versus density, as estimated from the compressibility of a variety of hydrocarbons. The initial peak locations are at 27 000 cm^{-1} for anthracene, 21 000 cm^{-1} for tetracene and 17 000 cm^{-1} for pentacene. The significant feature is that the slope increases rather rapidly with density. The ground state of these molecules is non-polar, while the excited state has a dipole moment. (The charge distribution is illustrated in Fig. 2.2 for anthracene.) The red shift is consistent with a higher moment after excitation. However, in the dipole approximation the energy change should be proportional to $1/R^3$, i.e., to the density. The observed change would seem to indicate that the charge separation increases with pressure. This is consistent with the greatly decreased electrical resistivity at high pressure. It is possible also that quadrupole terms are significant. Unfortunately, at the time these data were taken it was not possible to measure half widths or areas under the peak with any accuracy, and so to calculate thermal occupation of the π^* state.

The data for azulene contrast markedly, as shown in Figs. 6.18 and 6.19. In the low pressure region there is a distinct shift to higher energy with pressure, a maximum near 50 to 60 kbar, and a sharp decrease at higher pressures. The ground state of azulene has a dipole moment, so that there is evidently a decrease in moment upon excitation (in this

[*Refs. on p. 108*]

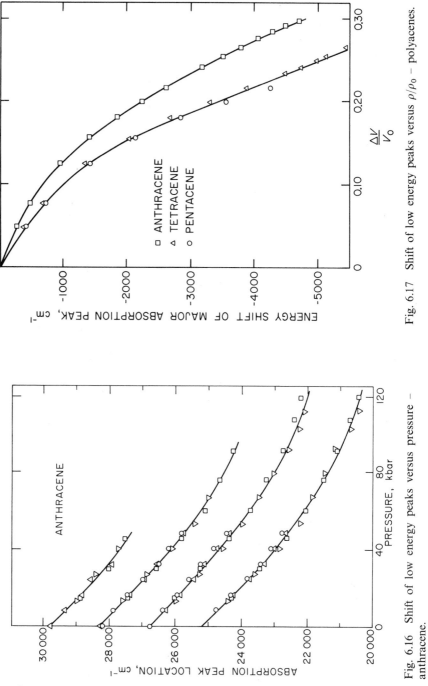

Fig. 6.17 Shift of low energy peaks versus ρ/ρ_0 – polyacenes.

Fig. 6.16 Shift of low energy peaks versus pressure – anthracene.

[Refs. on p. 108]

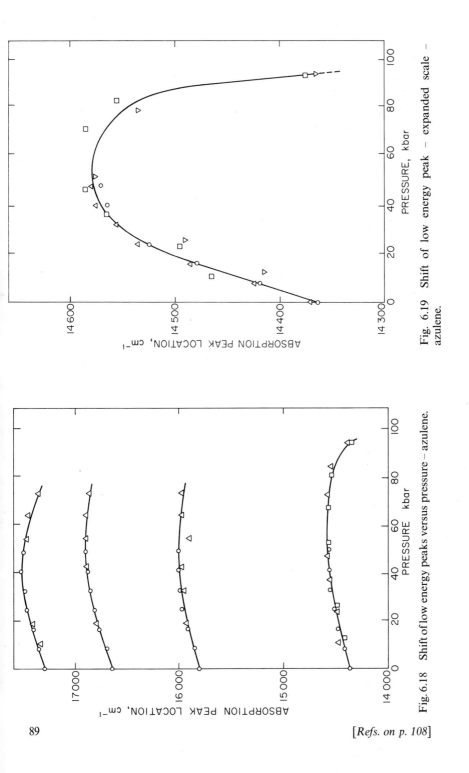

Fig. 6.19 Shift of low energy peak – expanded scale – azulene.

Fig. 6.18 Shift of low energy peaks versus pressure – azulene.

[*Refs. on p. 108*]

case to the ^1L_b state). As the pressure increases, the dipole moment of the excited state increases because of increasing charge separation, so that at sufficiently high pressure one obtains the decrease in transition energy which accompanies an increase in dipole moment upon excitation.

In the course of studies of the effect of pressure on the electronic structure and the oxidation state of iron in β-diketonate complexes, high pressure optical absorption studies were made [12] covering both the two charge transfer peaks in the regions 19 to 23 000 cm^{-1}, and 25 to 28 000 cm^{-1} and the $\pi \rightarrow \pi^*$ transitions lying near 28 to 35 000 cm^{-1}. The charge transfer transitions are discussed in connection with the reduction in Chapter 9. In Fig. 6.20 we plot the shift of the maxima in the π-π^* peaks for five compounds. (The numbering system refers to the tabulation in Fig. 6.10 above.) In all cases the shift is to lower energy. Fig. 6.21 exhibits the fractional change in peak half width with pressure.

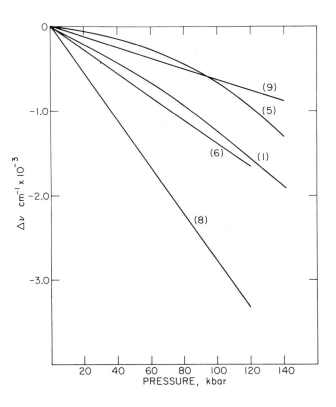

Fig. 6.20 Shift of π-π^* transition energy with pressure – five β-diketones.

[*Refs. on p. 108*]

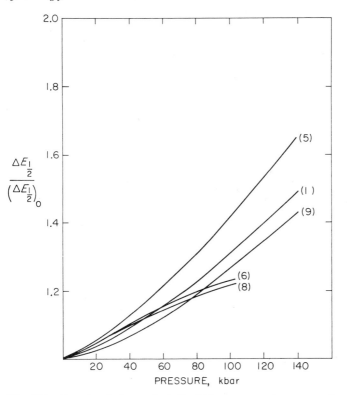

Fig. 6.21 Fractional change in half width versus pressure – π-π^* peaks in five β-diketones.

In Chapter 3 we developed equations relating half width and peak shift. From Equations (3.23) to (3.25) we see that

$$\frac{\partial h\nu_{\text{max}}}{\partial p} \bigg/ \left(\frac{\partial h\nu_{\text{max}}}{\partial p}\right)_0 = 1 + p\frac{\tau}{\Delta} = 1 + \frac{1}{(\delta E_{1/2})_0}\frac{\partial \delta E_{1/2}}{\partial p}.$$

Thus, if there is a large increase in half width with pressure, there should be a large increase in the magnitude of the slope of ν_{max} against pressure. This is precisely what is observed for compound TFACA(5). Compounds ACA(1) and MACA(9) exhibit an intermediate fractional change of half width with pressure, and a moderate increase in slope of the plot of $h\nu_{\text{max}}$ against pressure. Compounds FTFA(6) and BTFA(8) exhibit modest increases in half width and only slight increases in the slope of $h\nu_{\text{max}}$ against pressure. The results are certainly qualitatively consistent with the analysis of Chapter 3.

The phthalocyanines and porphyrins are frequently used as biological

[*Refs. on p. 108*]

prototype molecules. Their π-electron systems have been extensively studied. In later chapters we discuss changes of spin state and changes of oxidation state in ferrous phthalocyanine and ferric porphyrin derivatives. Here we discuss only certain π-π^* transitions for which we compare changes of peak location and half width with order of magnitude changes in intensity. One observes from Equation (3.22) of Chapter 3 that there is a relationship between the maximum in the optical absorption v_{max}, the peak half width, and the difference in thermal energy between, say, the π and π^* states. In particular, if a peak shifts sufficiently to lower energy and broadens sufficiently (particularly if it was initially broad), there is the possibility of partial occupation of the π^* orbital by π electrons. This could be promoted by configuration interaction and by other considerations outlined in Chapter 3. In this regard, we shall observe a difference in the behavior of the phthalocyanines and porphyrins.

Fig. 6.22 shows the structure of the phthalocyanine molecule. Four

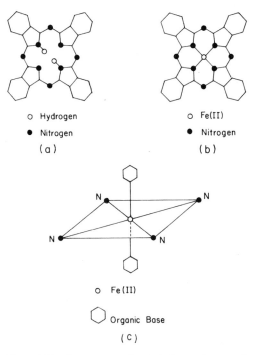

Fig. 6.22 Structure of phthalocyanine and derivatives: (a) phthalocyanine, (b) ferrous phthalocyanine, (c) substituted ferrous phthalocyanine.

[*Refs. on p. 108*]

pyrrole rings are bridged by nitrogens. Each pyrrole has a benzene ring fused to it. A metal atom may be placed at the center in square planar symmetry in the four pyrrole nitrogens. The metal ion used in the work discussed in this monograph is Fe(II). It is also possible to co-ordinate organic bases, such as pyridine, axially to the ferrous iron. In all cases the symmetry of the molecules is D_{4h}. Here we discuss the low energy π-π^* transition in phthalocyanine (Pc) and in the pyridine derivative of ferrous phthalocyanine FePc(pyr)$_2$ [26]. Figures 6.23 and 6.24 show the visible spectra for these two molecules in the solid state as a function of pressure. Molecular orbital calculations of Försterling and Kuhn [25] indicate that the highest occupied π level has a_{1u} symmetry. The wave function associated with this symmetry has maximum density on the pyrrole carbons nearest the nitrogens. The lowest $\pi \rightarrow \pi^*$ transition is $a_{1u} \rightarrow e_g^*$. The antibonding wave functions associated with the e_g^* orbitals have maximum density on the outer carbons of the pyrrole rings and on the nitrogen bridge atoms. Thus, the low energy transition discussed here involves some transfer of charge away from the center of the molecule.

As can be seen from Figs. 6.23 and 6.24, these peaks shift to lower energy with increasing pressure. In Pc they do not change significantly in width – in FePc(pyr)$_2$ they broaden somewhat –·but in both cases they are relatively narrow at all pressures. The intensities are established by normalization to the high energy peaks. Over 140 kbar they may lose a factor of about two in the intensity as estimated from peak areas.

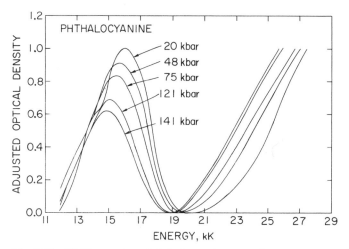

Fig. 6.23   Visible spectra of phthalocyanine at various pressures.

[*Refs. on p. 108*]

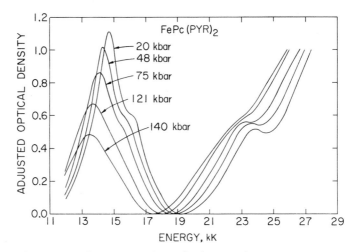

Fig. 6.24   Visible spectra of FePc(pyr)$_2$ at various pressures.

In the original paper [26] this loss was attributed to changes in the degree of configuration interaction among π orbitals with pressure. In Table 6.1 we calculate the energy E_{th} of the π^* state relative to the π state. It remains large and positive at all pressures. One must regard the calculation as approximate, especially since it was necessary to assume $\omega \cong \omega'$. Also, the factors discussed in Chapter 3 may act to reduce this difference. Nevertheless, the probability of thermal occupation of the π^* state seems slight, and the reasons originally assigned for the intensity change seem valid. However the $a_{2u} \rightarrow e_g^*$ transition near 3.75 eV has an initial half width of 0.9 eV. With pressure it shifts red and broadens noticeably, so that at pressures of about 75 to 80 kbar there is a definite probability of thermal occupation of the e_g^* state through this transition.

The structures of protoporphyrin IX and of some ferric derivatives are shown in Fig. 6.25. Here the pyrrole rings are joined by CH_2 groups, and various organic groups are attached to the periphery. The iron is again co-ordinated to four nitrogens. In hemoglobin four ferrous porphyrins are attached to the protein through imidazole groups. The sixth co-ordination site on the iron holds the oxygen in a manner not fully understood. In the free crystalline porphyrin under usual conditions, the iron is ferric iron. High pressure optical studies have been made on protoporphyrin (IX) (PRO), on imidazole protoheme (IMPH), wherein the iron is axially co-ordinated to two imidazoles, and on hemin and hematin. In hemin the iron is co-ordinated to a Cl^- and in hematin to

[*Refs. on p. 108*]

Table 6.1 Calculation of thermal energy difference between π^* and π orbitals

Pressure, kbar	$h\nu_{max}$ eV	$\Delta E_{1/2}$ eV	E_{th} eV
(a) Phthalcyanine (Pc)			
10	2.02	0.30	1.69
50	1.98	0.28	1.70
100	1.92	0.26	1.67
150	1.87	0.26	1.64
(b) Ferrous phthalocyanine dipyridine [FePc(pyr)$_2$]			
10	1.84	0.22	1.66
50	1.80	0.28	1.52
100	1.74	0.36	1.27
150	1.69	0.40	1.11
(c) Protoporphyrin IX (PRO)			
10	3.10	0.70	1.34
50	3.04	0.76	0.96
100	2.96	0.90	0.04
150	2.90	1.05	−1.08
(d) Imidazole protoheme (IMPH)			
10	2.98	0.50	2.08
50	2.95	0.75	0.93
100	2.90	0.92	−0.14
150	2.85	1.08	−1.23
(e) Hemin			
10	3.12	0.72	1.26
50	3.07	0.83	0.59
100	3.03	0.97	−0.35
150	2.99	1.17	−1.93
(f) Hematin			
10	3.15	0.82	0.73
50	3.09	0.96	−0.23
100	3.04	1.10	−1.31
150	3.00	1.15	−1.76

an OH$^-$. In these cases the sixth site is open and the iron is ∼0.5 Å out of the plane of the porphyrin. The pressure effects on the oxidation state and spin state of the iron are discussed in Chapter 10.

Optical absorption data at several pressures for PRO, IMPH, and hemin are shown in Figs. 6.26, 6.27, and 6.28 [28]. Spectra of hematin are very much like those of hemin. The atmospheric optical spectra have been widely analyzed by Williams [29, 30], by Falk [31, 32], by

[*Refs. on p. 108*]

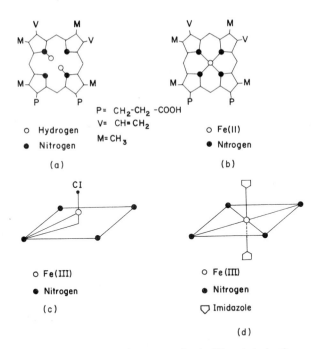

P= CH$_2$-CH$_2$ -COOH

V= CH=CH$_2$

M= CH$_3$

○ Hydrogen

● Nitrogen

(a)

○ Fe(II)

● Nitrogen

(b)

○ Fe(III)

● Nitrogen

(c)

○ Fe(III)

● Nitrogen

▽ Imidazole

(d)

Fig. 6.25 Structure of protoporphyrin IX and derivatives: (a) protoporphyrin IX, (b) protoheme IX, (c) hemin, (d) imidazole protohemichrome.

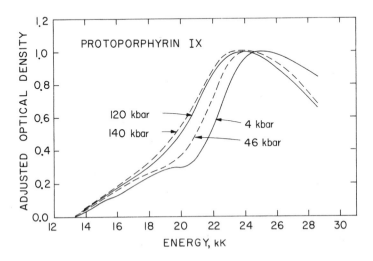

Fig. 6.26 Visible spectra of protoporphyrin IX at various pressures.

[Refs. on p. 108]

96

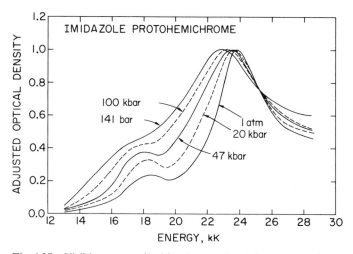

Fig. 6.27   Visible spectra of imidazole protohemichrome at various pressures.

Gouterman and co-workers [33–36], and by Harris [37, 38]. The π-π^* spectrum consists of low energy peaks of moderate intensity near $18\,000$ cm^{-1} (α and β) and the very intense Soret peak near $25\,000$ cm^{-1}. The α and β peaks shift to lower energy and fade rapidly with pressure. We concentrate on the behavior of the Soret peak. This peak is assigned to the $a_{1u} \rightarrow e_g^*$ transition as was the peak discussed in the phthalocyanines above. The a_{1u} orbital is concentrated on the carbons adjacent to the pyrrole nitrogens, while the e_g^* is again on the outer and bridge carbons,

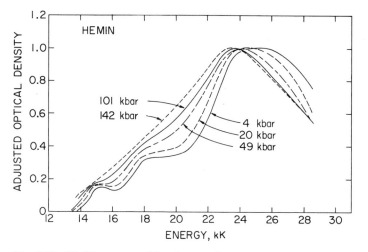

Fig. 6.28   Visible spectra of hemin at various pressures.

[*Refs. on p. 108*]

so that excitation involves moving electrons from the center towards the periphery. The peak shifts to lower energy and broadens rapidly with increasing pressure. While it is difficult to establish absolute intensities in the high pressure apparatus, the area under the Soret peak, based on various estimates decreases by a factor of 7 to 15 in 140 kbar. It is very doubtful whether configuration interaction alone can account for this large decrease.

Using Equation (3.22), a calculation of the probability of thermal occupation of the π^* state was performed for all three molecules. The results appear in Table 6.1. The calculations are very crude because of the assumptions discussed earlier and because of the asymmetric nature of the peaks. Half widths were measured to the low energy side of v_{max} and doubled. Nevertheless, there appears to be significant chance of thermal occupation of the π^* orbitals at high pressure. This probability is enhanced by the considerations discussed in Chapter 3. It appears quite reasonable that an important part of the loss in integrated intensity of the Soret peak may be caused by thermal occupation of the π^* level.

6.5 Charge transfer transitions

Electron transfer processes provide a wide variety of interesting chemistry. We discuss here pressure effects on optical peaks associated with two such processes: transfer between donors and acceptors in molecular complexes and transfer between metal and ligand in transition metal complexes. In later chapters we shall see how the shift in energy levels can lead to electronic transitions with chemical consequences.

As examples of molecular complexes, we discuss complexes of chloranil (tetrachloro-p-benzoquinone) and bromanil with durene (1,2,4,5 tetramethyl benzene), pentamethyl benzene (PMB), and hexamethyl benzene (HMB). Fig. 6.29 shows typical experimental results for HMB-chloranil [39]. The shift to lower energy occurs in many such complexes. Fig. 6.30 compares the complexes of chloranil with the three hydrocarbons. At 1 atm the peaks are located at 19 700, 20 100 and 19 000 cm^{-1} respectively for the durene, PMB, and HMB complexes in the solid state. We see that the complexes with HMB and PMB show relatively large shifts (2 200 to 2 400 cm^{-1} in 50 kbar) while the durene complex shifts significantly less.

The introduction of additional methyl groups to the benzene ring of the donor has two counteracting effects on the transition energy. The methyl groups lower the ionization potential of the donor and thus

[Refs. on p. 108]

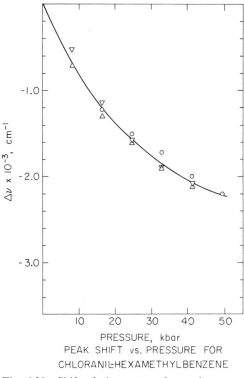

Fig. 6.29 Shift of charge transfer peak versus pressure – chloranil-hexamethylbenzene.

decrease the energy associated with the transition. However, they may also provide steric effects which tend to keep the donor and acceptor apart and thus increase the transition energy. It is possible that the odd number of methyl groups on PMB inhibits the packing and causes the relatively large atmospheric pressure transition energy. The lower ionization potential of the HMB may more than compensate for steric effects so that the transition energy of its complex at 1 atm is relatively low. If steric effects are present for the PMB and HMB complexes, the relatively large decrease in transition energy with pressure for these complexes may be a consequence of greatly increased overlap of donor and acceptor wave functions. The durene complex, with larger initial overlap, would exhibit a smaller change in transition energy. At 50 kbar the transition energies are approximately 18 400, 17 700 and 17 800 cm^{-1} for the 4, 5 and 6 methyl compounds respectively. The shifts may also be correlated with intensity changes as discussed below.

To obtain accurate data for the area under the absorption peak, and

[*Refs. on p. 108*]

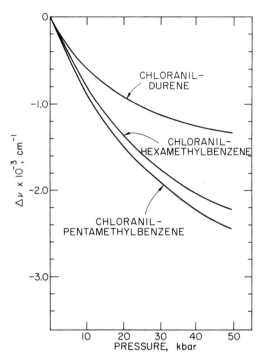

Fig. 6.30 Shift of charge transfer peak versus pressure – complexes of chloranil with three substituted benzenes.

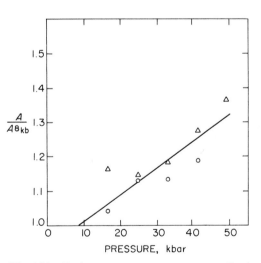

Fig. 6.31 Peak area versus pressure – normalized to 8 kbar – chloranil-hexamethylbenzene.

[*Refs. on p. 108*]

100

thus to establish transition probabilities, is quite difficult. In Fig. 6.31 we exhibit the estimated areas relative to the value at 8 kbar for chloranil-HMB. While the data are crude, one can estimate an increase of the order of 30% in 50 kbar. This is entirely in accord with the predictions of the theories discussed in Chapter 2. Similar data for the durene-chloranil complex indicate an intensity increase of about 45% in 50 kbar. If we now recall that according to theory the oscillator strength is given by the relationship

$$f \sim \nu \mu^2, \tag{6.4}$$

we see that the large red shift of the absorption peak for chloranil-HMB is qualitatively consistent with the relatively small increase in intensity, and vice versa for the durene complex. Of course, for many systems changes in the transition moment μ may be more important than changes in peak location.

Measurements of ν_{max} versus pressure were also made for the charge transfer peak in bromanil-HMB. The initial peak location is at 18 450 cm^{-1} in the solid. The shifts are compared with those for the chloranil complex in Fig. 6.32. Bromanil has a lower electron affinity than chloranil, but the stronger van der Waals attractive forces increase the overlap and compensate for the affinity.

If the transition dipole is independent of pressure, one should expect the shift of ν_{max} to be proportional to $1/R^3$, i.e., in the simplest order, proportional to $1/V$. Compressibility data are not available for the complexes; but, if one assumes that they are comparable with typical aromatic hydrocarbons, within the accuracy of the data a linear relationship between $1/V$ and ν_{max} exists [39].

The second type of charge transfer data to be discussed here is that between metal and ligand in transition metal complexes. The hexahalides of heavy transition metal ions offer an opportunity to relate pressure experiments to existing analysis. Jorgensen [40] has analyzed a number of these spectra. In particular, there are two relatively low lying peaks, or rather groups of peaks, lying near 16 to 18 000 cm^{-1} and 22 to 24 000 cm^{-1} for d^4 and d^5 ions. He has assigned these peaks to the transition π_L to t_{2g}, split by spin orbital coupling. In Figs. 6.33 and 6.34 we observe the shift of these peaks with pressure for $Na_2IrCl_6(5d^5)$ and $K_2OsBr_6(5d^4)$. [41]. We note that the center of the peak system shifts distinctly to lower energy with increasing pressure by ~ 2000 cm^{-1} in 100 kbar. The splitting increases drastically with pressure. At 1 atm it is about 5000 cm^{-1} for Na_2IrCl_6 and 5 500 cm^{-1} for K_2OsBr_6, while at 100 kbar

[*Refs. on p. 108*]

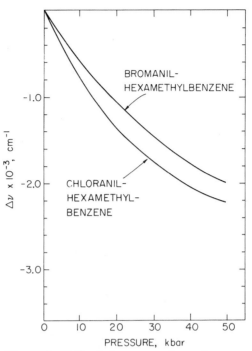

Fig. 6.32 Shift of charge transfer peak versus pressure – complexes of chloranil and bromanil with hexamethylbenzene.

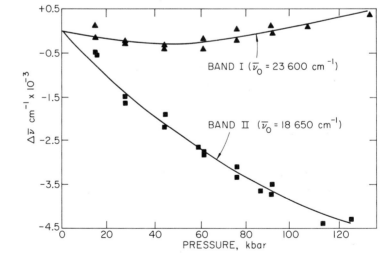

SHIFT OF CHARGE TRANSFER BAND MAXIMA WITH PRESSURE -Na_2IrCl_6

Fig. 6.33 Shift of metal to ligand charge transfer peak versus pressure – Na_2IrCl_6.

[Refs. on p. 108]

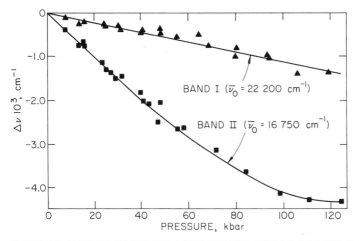

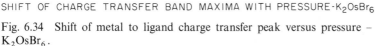

SHIFT OF CHARGE TRANSFER BAND MAXIMA WITH PRESSURE-K_2OsBr_6

Fig. 6.34   Shift of metal to ligand charge transfer peak versus pressure – K_2OsBr_6.

the splittings are $9\,000$ cm^{-1} and $8\,500$ cm^{-1}. The increased overlap and covalent interaction increases the probability of coupling between the metal electron and either of the nuclei. One may note that tetrahedral cobalt complexes [42] also exhibit significant increases in spin orbit coupling with pressure.

Other charge transfer peaks in metal-ligand systems are discussed in connection with the reduction of ferric iron in Chapters 9 and 10.

6.6   Absorption edges

A transition somewhat related to the charge transfer transitions discussed above can be observed in many insulating or semiconducting crystals whose electronic states can conveniently be described in terms of the band theory discussed in Chapter 2. As shown in Fig. 2.8, optical excitations can measure the energy differences between the top of the valence band and the bottom of the conduction band. These may be either direct transitions ($\delta\kappa = 0$) or indirect transitions with change in wave momentum. The effect of pressure can be to broaden the bands and decrease the gap, and can also involve a relative shift of one band with respect to the other which may tend either to increase or decrease the gap. For truly ionic crystals the valence band is made of anion wave functions while the conduction band will contain primarily cation wave functions. One might expect the conduction band, an excited state, to be

[*Refs. on p. 108*]

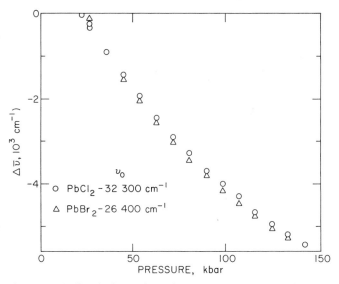

Fig. 6.35   Shift of absorption edge versus pressure PbCl$_2$ and PbBr$_2$ (V$_h^{16}$ structures).

more sensitive to compression than the valence band. One then might expect the shift of the edge to be relatively insensitive to the anion. Figure 6.35 shows this is indeed true for PbCl$_2$ and PbBr$_2$ [43]. Also, in Fig. 6.36 is plotted the absorption edge versus density for three thallous halides [43], all with the CsCl (simple cubic) structure. Again the same argument holds.

   Not all crystals exhibit such simple behavior with pressure. Fig. 6.37 shows the shifts of the absorption edge as a function of temperature and pressure for AgCl [44, 45]. In the low pressure (NaCl) phase there is a relatively small red shift of the edge with pressure which can be interpreted in terms of Seitz's suggestion that the tail of the edge in AgCl is due to an indirect transition, as shown in Fig. 2.8(c). At about 80 kbar there is a first-order phase transition accompanied by a large red shift of the edge. The high pressure phase also exhibits a small red shift of the edge with pressure.

   A consideration of the combined effect of temperature and pressure is useful. Normally, there is a large red shift of the absorption edge with pressure, and a modest red shift with increasing temperature. In AgCl, as can be seen from Fig. 6.37, there is a very large red shift (decrease in the energy gap) with increasing temperature at low pressure. This has been associated with a large increase in the number of Frenkel defects (lattice vacancies) with increasing temperature. As the pressure increases

[*Refs. on p. 108*]

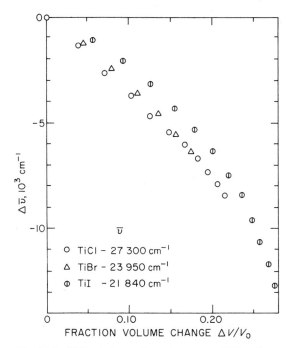

Fig. 6.36   Shift of absorption edge versus V/V_0 – three thallous halides.

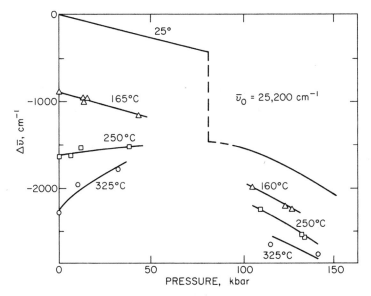

Fig. 6.37   Shift of absorption edge with pressure and temperature – AgCl.

[*Refs. on p. 108*]

the magnitude of the temperature coefficient decreases. Pressure evidently inhibits the formation of Frenkel defects. The net effect is then, an actual increase of the energy gap with increasing pressure along the 250 °C and 325 °C isotherms. In contrast, in the high pressure phase the behavior of the edge with increasing temperature is quite normal. Evidently lattice vacancies are never an important factor at these pressures.

Silicon, germanium, and the analogous III-V compounds having the zinc blende structure have been widely studied because of both their practical and theoretical interest. The measurement of their absorption edges as a function of pressure illustrates very nicely the differing effects of pressure on electronic levels of different symmetry.

Fig. 6.38 shows the shift of absorption edges of silicon and germanium [46] with pressure. Silicon exhibits a red shift of about 2×10^{-3} eV/kbar essentially independent of pressure. At low pressure the absorption edge of germanium shifts to a *higher* energy at a rate of about 7.5×10^{-3} eV/kbar. Around 30 to 35 kbar the direction of shift reverses and at high pressure germanium exhibits a red shift essentially identical with that of silicon.

Fig. 6.39 exhibits the corresponding data for GaSb [47]. Initially

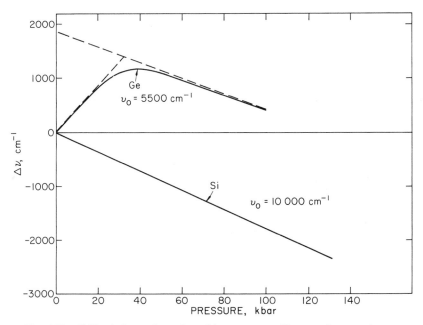

Fig. 6.38 Shift of absorption edge with pressure – silicon and germanium.

[*Refs. on p. 108*]

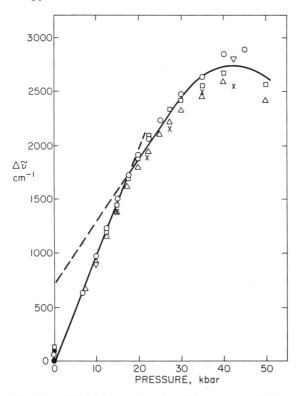

Fig. 6.39   Shift of absorption edge with pressure – GaSb.

there is a blue shift at a rate of 12×10^{-3} eV/kbar. In the region 15 to 20 kbar there is a distinct change in slope to about 7.3×10^{-3} eV/kbar. Around 50 kbar the slope changes sign, and at higher pressure there is evidence that there would be a red shift comparable to that observed for silicon.

These results can be explained in terms of a generalized plot of band energy versus wave vector as shown in Fig. 6.40. The symbols indicate directions and points of different symmetry in wave vector space. The notation can be found in books on band structure [48]. This exhibits all of the salient features of the band structure of this class of materials although it is not identical to any of them. It has been established by cyclotron resonance experiments that the transition observed in silicon is the indirect transition from $\kappa = 0$ in the valence band to the band minimum at Δ, in the 100 direction. In germanium the initial transition is to the band minimum in the 111 direction. In GaSb the transition observed at 1 atm is the direct transition at $\kappa = 0$. The 000 minimum

[*Refs. on p. 108*]

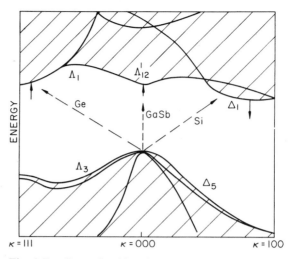

Fig. 6.40 Generalized band structure – Si, Ge, GaSb.

shifts to higher energy (relative to the valence band minimum) at a relatively rapid rate. The 111 minimum shifts to a higher energy at a somewhat lower rate, while the 100 minimum shifts to lower energy.

These results then illustrate the fact that electronic states of different symmetry are very differently affected by pressure.

References

1. S. MINOMURA and H. G. DRICKAMER, *J. Chem. Phys.*, **35** 903 (1961).
2. E. A. PEREZ-ALBUERNE and H. G. DRICKAMER, *J. Chem. Phys.*, **43** 1381 (1965).
3. H. G. DRICKAMER in *Solid State Physics*, Vol. 17, edited by F. Seitz and D. Turnbull, Academic Press, New York (1965).
4. H. HART and H. G. DRICKAMER, *J. Chem. Phys.*, **43** 2265 (1965).
5. H. G. DRICKAMER, *J. Chem. Phys.*, **47** 1880 (1967).
6. D. R. STEPHENS and H. G. DRICKAMER, *J. Chem. Phys.*, **34** 937 (1961).
7. J. HUBBARD, D. E. RIMMER and F. R. A. HOPGOOD, *Proc. Phys. Soc. (London)*, **88** 13 (1966).
8. J. C. ZAHNER and H. G. DRICKAMER, *J. Chem. Phys.*, **35** 1483 (1961).
9. P. J. WANG, M.S. Thesis, University of Illinois (1971).
10. M. PAGANNONE and H. G. DRICKAMER, *J. Chem. Phys.*, **43** 4064 (1965).
11. N. E. ERICKSON in *Mössbauer Effect and Its Applications in Chemistry*, edited by R. F. Gould *Amer. Chem. Soc., Washington* (1967).
12. C. W. FRANK and H. G. DRICKAMER, *J. Chem. Phys.*, **56** 3551 (1972).
13. J. H. SHULMAN and W. D. COMPTON, *Color Centers in Solids*, McMillan, New York (1962).
14. W. B. FOWLER, ed., *Physics of Color Centers*, Academic Press, New York, (1968).
15. E. MOLLWO, *Nachr. Ges. Wiss. Göttingen.*, Fachgruppen I, 97 (1931).

16. H. F. IVEY, *Phys. Rev.*, **72** 341 (1947).
17. I. S. JACOBS, *Phys. Rev.*, **93** 993 (1954).
18. W. G. MAISCH and H. G. DRICKAMER, *J. Phys. Chem. Solids*, **5** 328 (1958).
19. R. A. EPPLER and H. G. DRICKAMER, *J. Chem. Phys.*, **32** 1428 (1960).
20. P. W. BRIDGMAN, *Proc. Amer. Acad. Arts. Sci.*, **76** 1 (1945).
21. P. W. BRIDGMAN, *Proc. Amer. Acad. Arts. Sci.*, **71** 387 (1937).
22. H. OFFEN in *Organic Molecular Photophysics*, edited by J. B. Birks, Wiley-Interscience, New York (in press).
23. S. WIEDERHORN and H. G. DRICKAMER, *J. Phys. Chem. Solids*, **9** 330 (1959).
24. G. SAMARA and H. G. DRICKAMER, *J. Chem, Phys.*, **37** 474 (1962).
25. H. B. KLEVENS and J. R. PLATT, *J. Chem. Phys.*, **17** 470 (1949).
26. D. C. GRENOBLE and H. G. DRICKAMER, *J. Chem. Phys.*, **55** 1624 (1971).
27. H. D. FORSTERLING and H. KÜHN, *Intern. J. Quantum Chem.*, **2** 413 (1968).
28. D. C. GRENOBLE, C. W. FRANK, C. B. BARGERON and H. G. DRICKAMER, *J. Chem. Phys.*, **55** 1633 (1971).
29. R. J. P. WILLIAMS, *Chem. Rev.*, **56** 299 (1956).
30. R. J. P. WILLIAMS and D. W. SMITH, *Structure and Bonding*, **7** 1 (1970).
31. J. E. FALK and D. D. PERRIN in *Hematin Enzymes*, p. 56 edited by J. E. Falk, R. Lember and F. K. Morton, Pergamon Press, London (1961).
32. J. E. FALK, J. N. PHILLIPS and E. A. MAGNUSSON, *Nature*, **212** 1531 (1965).
33. C. WEISS, H. KOBAYASHI and M. GOUTERMAN, *J. Mol. Spect.*, **16** 415 (1965).
34. M. GOUTERMAN, *J. Chem. Phys.*, **30** 1139 (1959).
35. M. GOUTERMAN, *J. Mol. Spect.*, **6** 138 (1961).
36. M. GOUTERMAN and G. H. WAGIERE, *J. Mol. Spect.*, **11** 108 (1963).
37. G. HARRIS, *J. Chem. Phys.*, **48** 2191 (1968).
38. G. HARRIS, *Theoret. Chim. Acta.*, **17** 34 (1970).
39. R. B. AUST, M.S. Thesis, University of Illinois (1962).
40. C. K. JORGENSEN, *Molec. Phys.*, **2** 309 (1959).
41. A. S. BALCHAN and H. G. DRICKAMER, *J. Chem. Phys.*, **35** 356 (1961).
42. D. R. STEPHENS and H. G. DRICKAMER, *J. Chem. Phys.*, **35** 429 (1961).
43. J. C. ZAHNER and H. G. DRICKAMER, *J. Phys. Chem. Solids*, **11** 92 (1959).
44. T. E. SLYKHOUSE and H. G. DRICKAMER, *J. Phys. Chem. Solids*, **7** 207 (1958).
45. A. S. BALCHAN and H. G. DRICKAMER, *J. Phys. Chem. Solids*, **19** 261 (1961).
46. T. E. SLYKHOUSE and H. G. DRICKAMER, *J. Phys. Chem. Solids*, **7** 210 (1958).
47. A. L. EDWARDS and H. G. DRICKAMER, *Phys. Rev.*, **122** 1149 (1961).
48. J. C. SLATER, *Quantum Theory of Molecules and Solids*, Vol. 2, McGraw-Hill, New York (1965).

CHAPTER SEVEN

Electronic Transitions in Metals and Insulator-Metal Transitions

In this chapter we take up transitions which change the electrical properties of metals or compounds by establishing a new ground state for the system. Changes in physical rather than chemical properties are involved in these transitions, in contrast to the events discussed in later chapters.

7.1 Alkali metals

Free alkali atoms are characterized by a single outer s electron. In lithium and sodium the orbitals are filled according to the straightforward 'aufbau' principle. In potassium the 3d orbitals are unfilled and the highest energy electron is in a 4s orbital. Again in rubidium the 4d orbitals are higher in energy than the 5s and are therefore empty. Cesium exhibits empty 5d and 4f orbitals, with the highest energy electron in the 6s orbital. In the metals also at 1 atm the conduction band has primarily s character.

At 1 atm all the alkali metals have the bcc structure. At about 20 kbar cesium transforms to an fcc structure. Rubidium has a phase transition near 60 kbar, presumably also to the fcc structure. At low temperatures (77K to 200K) potassium undergoes a very sluggish transition near 220 kbar, which may be analogous to the one discussed above for the heavier alkalis. Sodium has no known transition; lithium undergoes a transition of unknown character near 60 kbar. We are concerned here with the behavior of the three heaviest alkali metals. From atomic spectra and various approximations to the band structure of these metals, it appears that the 5d band lies slightly above the 6s band in cesium with a 4f band at still higher energies. In rubidium the energy difference between the 5d and 4s bands is significantly larger than for the corresponding levels of cesium. The 3d band in potassium must lie well above the 4s band.

Bridgman [1, 2] measured both the volume and resistance of cesium as a function of pressure to about 70 to 75 kbar. In addition to the bcc → fcc transition near 20 kbar, he found a volume discontinuity at about 40 kbar. This volume change was accompanied by a very sharp cusp in the resistance, as shown in Fig. 7.1. Sternheimer [3] made a theoretical analysis which indicated that at this pressure the conduction band would take on predominantly d character (an electronic transition). While his analysis was approximate because of the boundary conditions used, his conclusions have not been challenged. There has been considerable further experimental work on this transition. Hall, Merrill, and

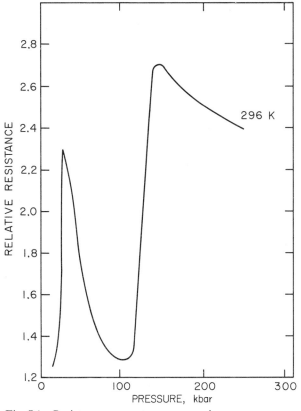

Fig. 7.1   Resistance versus pressure – cesium.

[*Refs. on p. 125*]

Barnett [4] showed that there were actually two closely spaced volume discontinuities near 40 kbar, but the structure remains fcc. Jayaraman *et al.* [5] showed that the resistance anomaly persists even above the melting point, so that electronic transitions can occur in liquids as well as solids.

When a higher pressure range became available, it was shown that the resistance exhibited a minimum near 100 kbar [6] and then rose to another sharp maximum near 135 kbar, as shown in Fig. 7.1. The nature of this event has not yet been thoroughly explored. It has been suggested that it involves a broadening of the 4f band and a lowering of its energy such that the conduction band may assume significant, if not predominant, 4f character, possibly with some 5p admixture, i.e., that a second electronic transition is involved. Wittig [7] has shown that the phase which appears above 100 kbar is superconducting. This is the first superconducting alkali metal. At one time it was considered that superconductivity was, in principle, impossible for these metals.

The electrical resistance of rubidium [8] as a function of pressure is shown in Fig. 7.2. The first significant feature is the sharp rise near 145 kbar at 296K. It is very probable that this is associated with a change from s to d character of the conduction band, as discussed above for cesium. At about 300 kbar there is a maximum in the resistance. It is very doubtful whether the 4f band could ever be lowered sufficiently with pressure to contribute significantly to the electronic properties. Some further rearrangement of the d band may be indicated, along with significant 4p hybridization.

The electrical behavior of potassium is shown in Fig. 7.3. The important feature for this discussion is the very sharp discontinuous rise in resistance near 260 kbar at 77K. The magnitude of this rise decreases with increasing temperature and it disappears near 250K, so that at room temperature the resistance increases with pressure with no discontinuities, as shown in Fig. 7.3. No serious mathematical analysis of the band structure of potassium as a function of pressure has yet been presented. It would seem quite reasonable to associate the discontinuity with an electronic transition, as typical diffusion-controlled transitions involving structure changes are invariably very sluggish at 77K. It is consistent with the previous discussion to associate it with a change from 4s to 3d character of the conduction band. It is possible that this transition disappears in a critical point. This has been demonstrated for the electronic transition in cerium discussed in the next section. It should be noted that the very sluggish transition at 220 kbar seems

[*Refs. on p. 125*]

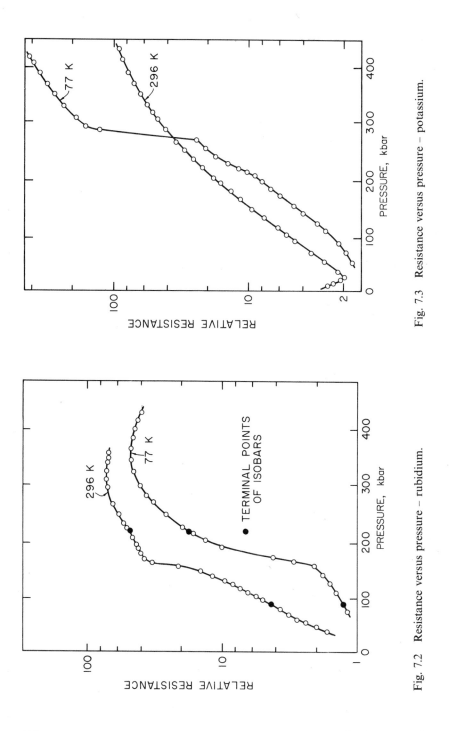

Fig. 7.3 Resistance versus pressure – potassium.

Fig. 7.2 Resistance versus pressure – rubidium.

113

[Refs. on p. 125]

also to disappear as temperature increases. If this is the bcc → fcc transition, it may be that the electronic transition will only occur in the fcc phase. The order of the transition pressures, K > Rb > Cs, is consistent with the relative energies of the nd orbitals *vis-à-vis* the $(n + 1)$ s orbitals in the three materials, as well as with the relative compressibilities.

7.2   Rare earth metals

The earliest identification of an electronic transition involved the rare earth metal cerium [9–11]. A discontinuity in resistance and in volume with no change in structure was identified as a change in the ground state by transfer of a 4f electron to the 5d band. More recently, theoretical work by Falicov [12] indicates that the actual transfer is probably from the 4f orbitals to the conduction band. Jayaraman [13] has followed the transition as a function of pressure and temperature and has shown that the discontinuity in resistance decreases with increasing temperature and disappears in a critical point. The isotherms of Fig. 7.4, taken from his work, illustrate the point.

Since the rare earth metals apparently involve orbitals or bands not too widely spaced in energy, electronic transitions of the above type should not be uncommon. Indeed, most of the rare earth elements show unusual resistance behavior with pressure [14]. Fig. 7.5 shows the higher pressure data for cerium. There is a second discontinuity near 65 kbar which may be electronic in origin. In Fig. 7.6 we see the resistance-pressure measurements for praseodymium. There are some low pressure events which may be electronic in character, but the major feature is the resistance maximum near 260 kbar, which sharpens significantly with decreasing temperature. Such behavior is much more characteristic of an electronic transition that of a diffusion-controlled phase change. Similar events have been observed in other rare earths.

Another type of electronic transition may occur in europium. The seven 4f electrons provide a half-filled shell – a particularly stable configuration. Fig. 7.7 illustrates the resistance-pressure behavior. In Fig. 7.8 we exhibit the results for barium, an alkaline earth metal. The similarities, particularly in the region 120 to 180 kbar, are striking. There are some speculations that the resistance discontinuity in barium at 120 kbar may involve a change from s to p character in the conduction band. A similar transition in europium seems particularly feasible in view of the stability of its 4f configuration.

[*Refs. on p. 125*]

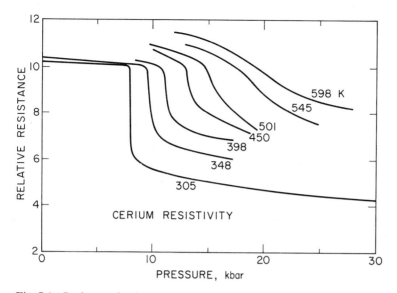

Fig. 7.4   Resistance isotherms – low pressure region – cerium.

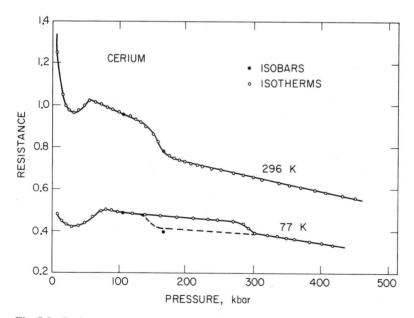

Fig. 7.5   Resistance versus pressure – cerium.

115

[*Refs. on p. 125*]

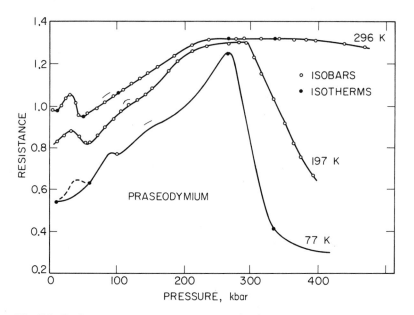

Fig. 7.6   Resistance versus pressure – praseodymium.

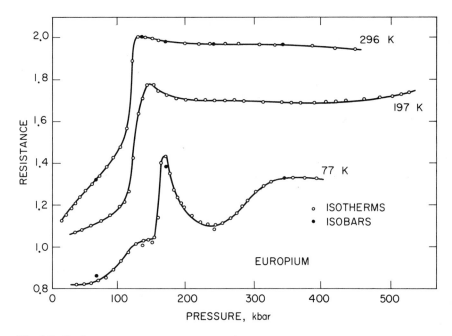

Fig. 7.7   Resistance versus pressure – europium.

[*Refs. on p. 125*]   116

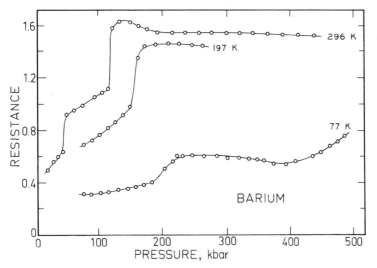

Fig. 7.8   Resistance versus pressure – barium.

7.3   Calcium, strontium and ytterbium

The alkaline earth metals present something of an anomaly. The free atoms consist only of filled or empty sets of orbitals so that, in the simplest view, a solid array of such atoms should be an insulator or semiconductor. These materials are metals because the conduction band

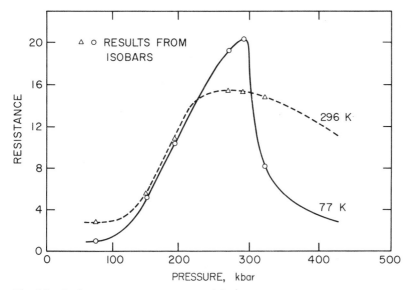

Fig. 7.9   Resistance versus pressure – calcium.

[*Refs. on p. 125*]

is a complex hybrid of s, p, and d orbitals so that the intersection
of the Fermi surface with the Brillouin zone boundary is complex.

The alkaline earth metals, calcium and strontium, and the closely
related metal ytterbium exhibit unusual behavior [15, 16]. Fig. 7.9
illustrates this for calcium. At low pressures it exhibits metallic behavior.
Somewhere below 200 kbar the isotherms cross, and at higher pressures
the resistance decreases with increasing temperature. Isobars indicate
that the decrease is at least approximately exponential. Near 300 kbar
there is apparently a first-order phase transition, and at higher pressures
the behavior is again metallic.

Strontium exhibits analogous behavior at much lower pressures. The
metallic behavior disappears near 20 kbar and near 40 kbar there is a
first-order phase transition from fcc to bcc, with metallic behavior at
higher pressures. Fig. 7.10 exhibits the very similar behavior of ytterbium,
which has a filled 4f shell.

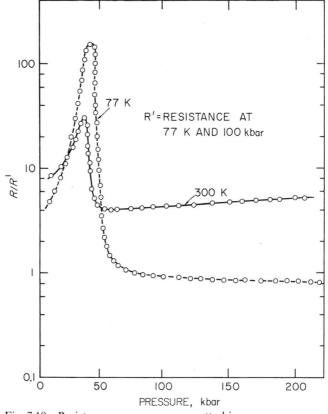

Fig. 7.10   Resistance versus pressure – ytterbium.

[*Refs. on p. 125*]

There have been extensive theoretical analyses of calcium. Cracknell [17] reviews the theory to 1969. Calcium, and probably also strontium, become semimetals rather than semiconductors. A recent study by McCaffrey *et al.* [18] is also consistent with this viewpoint. McWhan, Rice and Schmitt [19] have made a particularly thorough experimental study and analysis for ytterbium. Although the resistance behavior is very similar to that of calcium and strontium, it becomes a semiconductor rather than a semimetal. Very small variations in the relationship of the Fermi surface to the Brillouin zone boundary are sufficient to account for this difference. Important contributions to the study of ytterbium and strontium have also been made by Hall and Merrill [20], by Souers and Jura [21], by McWhan and Jayaraman [22], and by Jerome and Rieux [23].

7.4   Insulator-metal transitions

From the earliest days of band theory, as in the pioneering text of Seitz [24], it has been illustrated that decreased interatomic distance should broaden energy bands. Ultimately, this broadening should result in overlap between a filled valence band and an empty conduction band, and thus a possibly continuous transition from insulator to metal. It is, of course, difficult to separate the result of this broadening from relative shifts of the center of gravity of the bands which may enhance or impede the first effect. Data for the molecular crystal iodine illustrate the general process [25, 26]. Iodine crystallizes in a base-centered orthorhombic structure with the I_2 molecules in the *ac* plane. In Fig. 7.11 is plotted resistance (on a logarithmic scale) versus pressure. Depending on the orientation of the leads, the resistance can be measured either in the *ac* plane, or perpendicular to it. The resistance drops by many orders of magnitude (estimated resistivity at 200 kbar is 10^{-2} to 10^{-3} Ω cm). For measurements in the *ac* plane, there is a distinct break in slope near 170 kbar; a similar break occurs near 130 kbar for resistance perpendicular to the *ac* plane. Above the break in slope the resistance continues to decrease, but at a rate more typical of metals. In Fig. 7.12 we plot the energy gap obtained both from optical measurements of the absorption edge and from the temperature coefficient of electrical resistance through Equation (2.20) of Chapter 2. The agreement between the two measurements is excellent – probably fortuitously so. Up to about 100 kbar, E_g is independent of direction. The energy gap goes to zero for measurements in the *b*

[*Refs. on p. 125*]

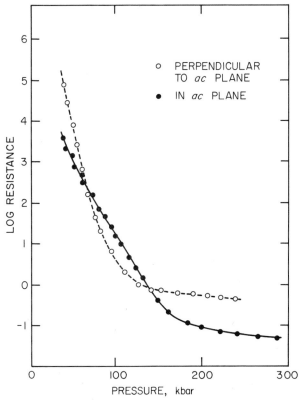

Fig. 7.11   Log resistance versus pressure – iodine.

direction (perpendicular to the *ac* plane) near 130 kbar and for measurements in the *ac* plane near 170 kbar; in each case, where the breaks in slope of the resistance – pressure curves occur. As can be seen in Fig. 7.13, the resistance at high pressure increases nearly linearly with temperature, as would be expected for a metal. Between 130 and 170 kbar iodine apparently exhibits metallic behavior in one direction and semiconducting behavior in the other, much like single crystal graphite at 1 atm. No discontinuity in resistance was observed at the transition at temperatures down to 77K. The possibility of a small discontinuity cannot be precluded because of the rapid change of resistance with pressure. X-ray diffraction measurements [27], while crude, indicate no change in structure or volume discontinuity to well above 200 kbar. Similar behavior, although less completely documented, has been observed for the thallous halides [28]. Certain aromatic hydrocarbons also become metallic at high pressure. These are discussed briefly in Chapter 11.

[*Refs. on p. 125*]

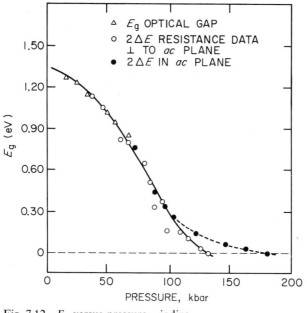

Fig. 7.12 E_g versus pressure – iodine.

A second mechanism whereby an insulator or semiconductor might become a metal would involve the rearrangement of the atoms or ions into a new lattice wherein there might be appropriate overlap of filled and empty bands. Relatively open structures are obvious candidates. The transformation with temperature from grey tin, which has the cubic diamond structure and is semiconducting, to white tin, which is a tetragonal metal, has been known for decades. Such transformations have been observed as a function of pressure in silicon and germanium [29], as shown in Fig. 7.14. One sees a large discontinuity in resistance in germanium near 105 to 110 kbar, and in silicon near 145 to 150 kbar. The high pressure phase is metallic and has the white tin structure, as shown by Jamieson [30]. For materials like these, it is difficult to establish a true equilibrium pressure for the transition, as metastability is always a serious problem. Also Bundy and Kasper [31] have been able to quench in from high pressure phases of silicon and germanium which are neither diamond nor white tin. Nevertheless, Van Vechten [32] has successfully calculated the transition pressures for silicon and germanium from a general model, with good agreement with the experimental observations. Similar semiconductor-metal transitions accompanied by structure changes occur in a large number of III-V and

[*Refs. on p. 125*]

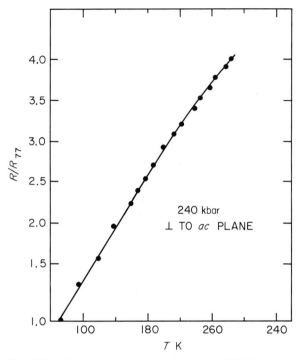

Fig. 7.13   Resistance versus temperature at 240 kbar –
iodine.

II-VI compounds where the low pressure phase has the zinc blende
analog of the diamond structure. Fig. 7.15 shows results for ZnS, ZnSe,
and ZnTe [33]. For a number of these it has been shown that the high
pressure material has a structure which is the diatomic analog of white
tin.

As mentioned in Chapter 2, there is a third form of insulator-metal
transition which is more clearly an 'electronic transition' within the
framework of our definition than are the first two. Transition metal
compounds such as NiO or V_2O_3 should, in principle, be metals because
there should be a partially filled energy band arising from the atomic
d orbitals which are not filled. Many of these are insulators, although
there exist also oxide metals. Mott developed the theory outlined
briefly in Chapter 2, which involves a critical parameter a_h such that
above a critical electron density one obtains metallic conduction. There
exist tests of this theory involving increased doping of semiconductors
until metallic conduction obtains [34], but these are of peripheral interest
here. A number of transition metal oxides undergo insulator-metal

[*Refs. on p. 125*]

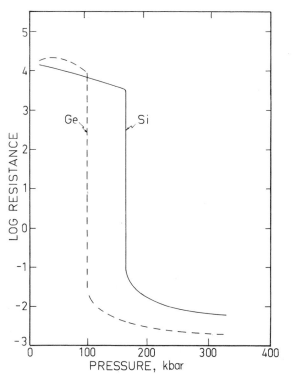

Fig. 7.14 Insulator-metal transitions – Si and Ge.

transitions with changing temperatures, but an analysis makes it doubtful whether these are indeed 'Mott transitions'.

It is clear that compression of the lattice could in principle bring about such a transition. There have been a variety of experiments on transition metal oxides with this test in mind. Probably the most interesting and informative have been those performed at Bell Laboratories by McWhan and his colleagues [35–38]. They have studied a number of different mixed oxides, in particular, crystals of the formula $V_{1-x}Cr_xO_3$. They have shown that the crystal $V_{0.96}Cr_{0.04}O_3$ exhibits most, if not all, the features of a Mott transition with regions of temperature and pressure where it is a paramagnetic insulator, an antiferromagnetic insulator, or a metal. Even for this material the transition may not be a pure 'electron-electron' interaction as demanded by the Mott theory, as it may be phonon-assisted. Nevertheless, they have developed a particularly interesting and informative class of materials. An especially important aspect of their work is the demonstration of the

[*Refs. on p. 125*]

equivalence of changing pressure and changing chemical composition on the electronic properties.

A fourth type of insulator-metal transition has recently been uncovered, which involves the samarium chalcogenides SmS, SmSe, and SmTe[39]. These divalent compounds of samarium are all semiconductors at 1 atm. With increasing pressure, the Sm^{2+} undergoes a Rydberg transition much like that discussed for the rare earth metals above. It

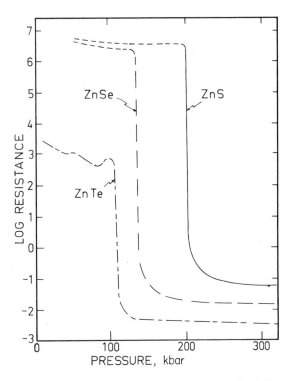

Fig. 7.15   Insulator-metal transitions – ZnS, ZnSe, and ZnTe.

donates an electron from the 4f level of the samarium to the conduction band of the compound which provides metallic conductivity. This event occurs discontinuously at 6.5 kbar and 25 °C in SmS, but in the heavier chalcogenides it occurs continuously over a considerable range of pressure. In the following chapters we shall see many examples of electronic transitions which occur over a range of pressure rather than discontinuously.

[*Refs. on p. 125*]

References

1. P. W. BRIDGMAN, *Proc. Amer. Acad. Arts and Sci.*, **76** 55 (1948).
2. P. W. BRIDGMAN, *Proc. Amer. Acad. Arts and Sci.*, **81** 165 (1952).
3. R. S. STERNHEIMER, *Phys. Rev.*, **78** 235 (1950).
4. H. T. HALL, L. MERRILL and J. D. BARNETT, *Science*, **146** 1297 (1964).
5. A. JAYARAMAN, R. C. NEWTON and J. M. MCDONOUGH, *Phys. Rev.*, **159** 527 (1967).
6. R. A. STAGER and H. G. DRICKAMER, *Phys. Rev. Letters*, **12** 19 (1964).
7. J. WITTIG, *Phys. Rev. Letters*, **24** 812 (1970).
8. R. A. STAGER and H. G. DRICKAMER, *Phys. Rev.*, **132** 124 (1963).
9. A. W. LAWSON and T. V. TANG, *Phys. Rev.*, **76** 301 (1949).
10. I. LIHKTER, N. RIABININ and L. F. VERESCHAGIN, *Soviet Phys. JETP*, **6** 469 (1958).
11. R. HERMAN and C. A. SWENSON, *J. Chem. Phys.*, **29** 398 (1958).
12. L. M. FALICOV and R. RAMIREZ, *Phys. Rev.*, **B3** 2425 (1971).
13. A. JAYARAMAN in *Physics of Solids at High Pressure*, p. 478, edited by C. T. Tomizuka and R. M. Emrick, Academic Press, New York (1965).
14. R. A. STAGER and H. G. DRICKAMER, *Phys. Rev.*, **133** 830 (1964).
15. R. A. STAGER and H. G. DRICKAMER, *Phys. Rev.*, **131** 2524 (1963).
16. R. A. STAGER and H. G. DRICKAMER, *Science*, **139** 1284 (1963).
17. A. P. CRACKNELL, *Adv. in Physics* **18** 681 (1969).
18. J. W. MCCAFFREY, D. A. PAPACONSTANTOPOULOS and J. R. ANDERSON, *Solid State Commun.*, **8** 2109 (1970).
19. D. B. MCWHAN, T. M. RICE and P. H. SCHMIDT, *Phys. Rev.*, **177** 1063 (1969).
20. H. T. HALL and L. MERRILL, *Inorg. Chem.*, **2** 618 (1963).
21. P. C. SOUERS and G. JURA, *Science*, **140** 481 (1963).
22. D. B. MCWHAN and A. JAYARAMAN, *App. Phys. Letters*, **3** 129 (1963).
23. D. JEROME and M. RIEUX in *Properties Physiques des Solides sous Pression*, edited by D. Bloch CNRS, Paris (1970).
24. F. SEITZ, *Modern Theory of Solids*, McGraw-Hill, New York (1940).
25. A. S. BALCHAN and H. G. DRICKAMER, *J. Chem. Phys.*, **34** 1948 (1961).
26. B. M. RIGGLEMAN and H. G. DRICKAMER, *J. Chem. Phys.*, **37** 446 (1962); **38** 2721 (1963).
27. R. W. LYNCH and H. G. DRICKAMER, *J. Chem. Phys.*, **45** 1020 (1966).
28. G. A. SAMARA and H. G. DRICKAMER, *J. Chem. Phys.*, **37** 408 (1962).
29. S. MINOMURA and H. G. DRICKAMER, *J. Phys. Chem., Solids*, **23** 451 (1962).
30. J. C. JAMIESON, *Science*, **139** 762, 845 (1963).
31. F. P. BUNDY and J. S. KASPER, *Science*, **139** 340 (1963).
32. J. A. VAN VECHTEN, *phys. stat. sol.*, **47** 261 (1971).
33. G. A. SAMARA and H. G. DRICKAMER, *J. Phys. Chem. Solids*, **23** 457 (1962).
34. M. N. ALEXANDER and D. F. HOLCOMB, *Rev. Mod. Phys.*, **40** 815 (1968).
35. D. B. MCWHAN, T. M. RICE and J. P. REMEIKA, *Phys. Rev. Letters*, **23** 1384 (1969).
36. D. B. MCWHAN and J. P. REMEIKA, *Phys. Rev.*, **B2** 3734 (1970).
37. A. JAYARAMAN, D. B. MCWHAN, J. P. REMEIKA and P. D. DERNIER, *Phys. Rev.*, **B2** 3751 (1970).
38. A. C. GOSSARD, D. B. MCWHAN and J. P. REMEIKA, *Phys. Rev.*, **B2** 3762 (1970).
39. A. JAYARAMAN, V. NARAYANAMURTI, E. BUCHER and R. G. MAINES, *Phys. Rev. Letters*, **25**, 368, 1430 (1970).

CHAPTER EIGHT

Spin Changes in Iron Complexes

As discussed in Chapter 2, the normal ground state of a free ion is the one of maximum multiplicity (Hund's First Rule). This configuration minimizes the interelectronic repulsion. It was also shown there that a ligand field of octahedral symmetry partially removes the degeneracy, giving a lower lying triplet of $\pi(t_{2g})$ symmetry and a higher energy doublet of $\sigma(e_g)$ symmetry. We showed in Chapter 6 that for high spin complexes the ligand field increases significantly with pressure, and that the interelectronic repulsion (Racah) parameters decrease. Thus, on the one hand, with pressure it takes an increasing amount of potential energy to occupy all the orbitals; on the other hand, the decreasing Racah parameters decrease the spin pairing energy. Both of these factors increase the probability of a high spin to low spin conversion. (The increase in field is probably the major effect.) Since Griffith [1] has shown that an intermediate spin configuration is never the most stable in octahedral symmetry, we need only deal with high and low spin states at this point.

8.1 High spin to low spin transformation

In Chapter 5 we demonstrated that for the ferrous ion the high spin and low spin states are identifiable by Mössbauer resonance. As an example of the high spin-low spin transition we discuss iron as a dilute substitutional impurity in MnS_2, a cubic crystal, isomorphous onto FeS_2 (iron pyrites). However, the lattice parameter of the former is

[Refs. on p. 151]

6.102 Å while that of the latter is 5.504 Å. It is therefore not surprising that, although Fe(II) in FeS_2 is low spin at all pressures [2], Fe(II) as a dilute substitutional impurity in MnS_2 is high spin at atmospheric pressure. Of course, there may be local relaxation near the Fe(II) so that the local lattice parameter near it may differ significantly from the bulk value. Nevertheless, it is reasonable to conceive of the iron in MnS_2 as under a large negative pressure (expanded lattice) compared with its situation in FeS_2. The majority of the pressure measurements have involved 2% ^{57}Fe in MnS_2 [3]. Some measurements with 0.5% impurity gave no important differences. The isomer shift (0.84 mm s^{-1} relative to bcc iron) and quadrupole splitting (1.50 mm s^{-1}) differ considerably from the typical high spin Fe(II) values observed in ionic compounds (see Chapter 6). They compare closely, however, with the values obtained for other high spin ferrous sulfides. There was usually a trace ($\sim 10\%$ or less) of material with isomer shift ~ 0.35 mm s^{-1} and quadrupole splitting ~ 0.60 mm s^{-1} which compares closely to the values for FeS_2 (IS = 0.30 mm s^{-1} and QS = 0.60 mm s^{-1}). This material may have been due to some clustering of the iron or to Fe(II) located at or near defects. A typical atmospheric pressure spectrum appears in the upper half of Fig. 8.1. As pressure was applied there was no significant change up to about 40 kbar. At higher pressures the relative amount of low spin Fe(II) increased, as shown in the lower half of Fig. 8.1, so that at 65 kbar the conversion was over 50%. In Fig. 8.2 it can be seen that by 138 kbar the conversion was complete. Fig. 8.3 plots conversion versus pressure. The process is reversible but involves considerable hysteresis. This is not surprising as the bonding and the volume associated with the two spin states are different. Similar transitions were observed for ^{57}Fe as an impurity in $MnSe_2$ and $MnTe_2$.

This is an example of an electronic transition from a paramagnetic to a diamagnetic ground state. The transition is brought about by the increasing field of the ligands with decreasing Fe(II) – ligand distance. In these crystals it occurs over a finite but modest range of pressures. There was no tendency for the conversion to change with time at a given pressure. It is not at present clear whether it would occur more nearly discontinuously in a perfect crystal in a hydrostatic medium. In contrast to other spin changes discussed later in this chapter, and changes in oxidation state discussed in later chapters, this transition goes to completion.

The spin change observed here should be of significance in geochemistry because pressures in the mantle reach 1500 kbar at the core

[*Refs. on p. 151*]

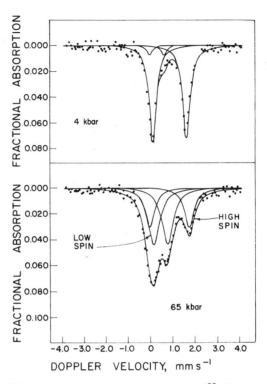

Fig. 8.1 Mössbauer spectra – $MnS_2(^{57}Fe)$ – 4 kbar and 65 kbar.

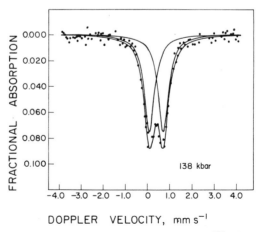

Fig. 8.2 Mössbauer spectra – $MnS_2(^{57}Fe)$ – 138 kbar.

[*Refs. on p. 151*]

128

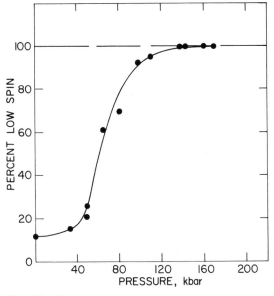

Fig. 8.3   Conversion to low spin versus pressure –
MnS$_2$(^{57}Fe).

boundary. It appears likely that ferrous ions, at least, should all be low
spin well before the core is reached.

8.2   Low spin to high spin transformations

The HS → LS transition discussed above has long been predicted from
theory and has an obvious source. Partial transformations from LS → HS
have also been observed with increasing pressure. This event is surprising, based both on electronic and thermodynamic considerations. As
discussed above, one normally expects an increase in ligand field with
pressure, reducing the probability of the high spin state. Also, one
normally associates a smaller volume (shorter bond distance) with a low
spin ion than with the corresponding high spin ion.

From the standpoint of thermodynamics, it must be remembered that
the criterion for a process to increase conversion with pressure is that
the volume of the system *as a whole* decrease with increasing conversion
at constant temperature and pressure. It is not necessary that *every* bond
shorten; some may shorten while others lengthen. Also, even if the bond
distances do not shorten, the distance between molecules or complexes in
the lattice may decrease because, in the transformed state, the attractive

[*Refs. on p. 151*]

intermolecular forces or the packing efficiency may be greater. In addition, one state may occupy a larger volume in the lattice than a second state at 1 atm, but it may be more compressible, so that at sufficiently high pressure the volumetric situation is inverted. This is one of the considerations treated in Chapter 4.

Before studying the experimental evidence and offering an explanation from a electronic viewpoint, it is useful to understand the bonding mechanism which is operative in most complexes which are low spin at 1 atm. In most cases the spin state is controlled by the degree of back donation. This phenomenon was mentioned in Chapter 2. In Fig. 8.4 we

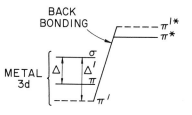

Fig. 8.4   Effect of back donation
on molecular orbitals.

illustrate the effect. The ligands typically have π^* orbitals which are empty, and which are not too high in energy above the π orbitals. They also have the appropriate symmetry to bond with the metal d_π orbitals. The metal ions then donate electrons into these orbitals (hence 'back donation'). Since the metal d_π orbitals are bonding in this context, they are stabilized with regard to their 'normal' state. (See the dotted lines in Fig. 8.4.)

The ligand field is thus increased from Δ to Δ' which may be sufficient to overcome the spin pairing energy and so to stabilize the low spin state. For Fe(II) the delocalization of the 3d electrons also reduces the shielding of the 3s electrons and provides the relatively small isomer shift characteristic of low spin Fe(II), as discussed in Chapter 5. In fact, Erickson [4] has shown that there is an excellent correlation between low spin isomer shift and the size of the ligand field. The greater the back donation, the larger the ligand field and the smaller the isomer shift. We may then expect that any changes from low to high
[*Refs. on p. 151*]

spin will be associated with changes in back donation, and therefore with changes in the availability of the π^* orbitals for occupation by metal electrons.

8.2.1 *Phenanthroline complexes*

The most complete study of these spin changes has involved complexes of phenanthroline [5, 6]. Complexes of iron with this molecule have been widely studied for the purpose of analytical chemistry; it also has some characteristics of a biological prototype. Phenanthroline is shown in Fig. 8.5. It is a planar heterocyclic aromatic molecule with nitrogens at the 1 and 10 positions. It is possible to substitute various electron donating or withdrawing groups at the other positions and thus to affect the bonding characteristics of the nitrogens [6, 7]. It is also possible to form two types of quasioctahedral complexes of phenanthroline with iron. In the tris complexes three phenanthrolines each occupy two of the co-ordination sites. The anions are then outside the co-ordination sphere. These complexes typically have ligand fields of 16 to 20 000 cm^{-1} and are low spin.

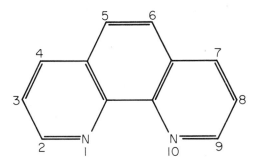

Fig. 8.5 Structure of phenanthroline.

Bis complexes involve two phenanthrolines co-ordinated to the iron, with anions occupying the other sites, usually in the cis configuration. The anions may be Cl^-, NCO^-, NCS^-, $NCSe^-$, N_2^-, CN^- etc. The properties of these complexes depend on the degree of back donation to the anion ligand. Generally they have ligand fields in the range 10 to 14 000 cm^{-1} and are high spin, although the bis cyanide is low spin. (Similar complexes with 2, 2' bipyridil were also studied in the original work [5]. Since they showed no qualitatively different effects they will not be discussed here.)

In Figs. 8.6 and 8.7 we show typical Mössbauer spectra for a tris complex (the chloride). At 3 kbar one observes only a typical low spin

[*Refs. on p. 151*]

[Refs. on p. 151]

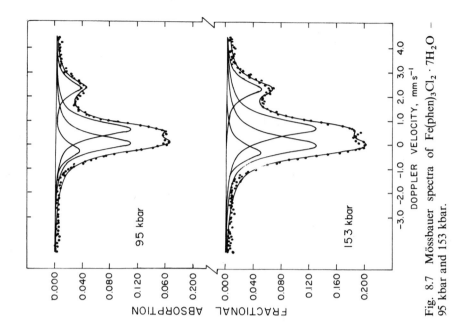

Fig. 8.7 Mössbauer spectra of Fe(phen)$_3$Cl$_2$ · 7H$_2$O –
95 kbar and 153 kbar.

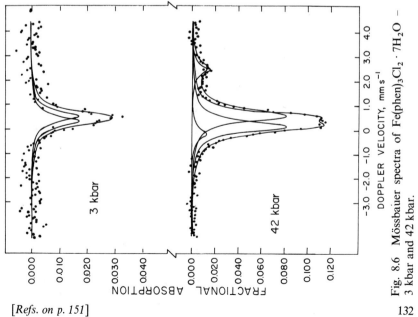

Fig. 8.6 Mössbauer spectra of Fe(phen)$_3$Cl$_2$ · 7H$_2$O –
3 kbar and 42 kbar.

Fe(II) spectrum (low IS and QS). At 42 kbar measurable high spin appears. By 95 kbar there is perhaps 20% high spin, and by 153 kbar there is 25%. Other tris complexes may show larger or smaller yields. Fig. 8.8 shows per cent low spin as a function of pressure for a series of tris complexes, at 298K. In general, the amount of low spin decreased with increasing temperature. For the tris thiocyanate, at 80 kbar and 298K there was 87% low spin, and at 335K 72% low spin.

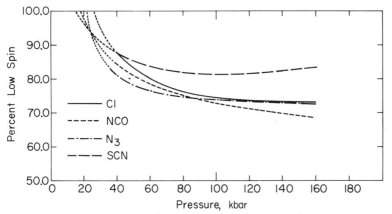

Fig. 8.8 Conversion to high spin versus pressure – ferrous tris phenanthroline complexes.

The bis complexes typically showed even more unusual behavior as exhibited for the bis isothiocyanate in Figs. 8.9 and 8.10. At 1 atm one has a characteristic high spin spectrum. (The IS $\simeq$ 1.05 mm s^{-1} is a little low for ionic high spin compounds, but the state is well established.) The ligand field is near the crossover point, so that by 15 kbar one has 75% low spin. At 32 kbar the conversion is over 80%, but at higher pressures the amount of low spin has distinctly dropped. Fig. 8.11 shows per cent low spin versus pressure for a series of compounds. The conversion from high to low spin depends on the size of the ligand field, i.e., the amount of back donation to the anion ligands. The degree to which the conversion reverses depends on the same factor. The chloride, which has no empty π orbitals, converts only at relatively high pressures and never reverses, although there is a distinct change in slope near 100 kbar.

The effects of substitution on the phenanthroline could be predicted from the electron withdrawing ability of the substituent. This is illustrated for three tris complexes substituted in the 5 position in Fig. 8.12. (5 Meph = CH_3^- in the 5 position, 5 Clph = Cl^- in the 5 position, and

[*Refs. on p. 151*]

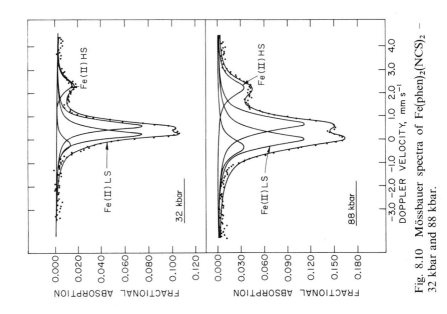

Fig. 8.10 Mössbauer spectra of Fe(phen)$_2$(NCS)$_2$ – 32 kbar and 88 kbar.

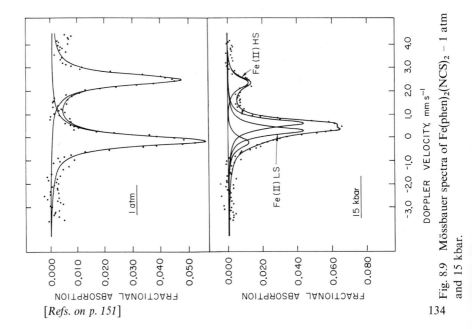

Fig. 8.9 Mössbauer spectra of Fe(phen)$_2$(NCS)$_2$ – 1 atm and 15 kbar.

[Refs. on p. 151]

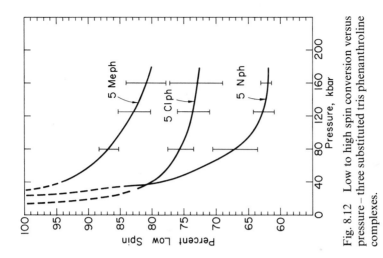

Fig. 8.12 Low to high spin conversion versus pressure – three substituted tris phenanthroline complexes.

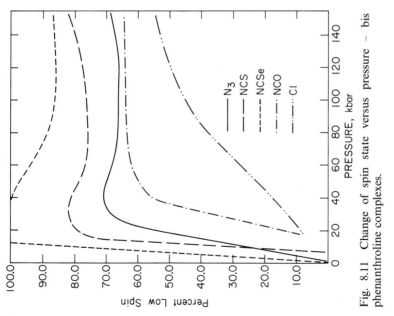

Fig. 8.11 Change of spin state versus pressure – bis phenanthroline complexes.

[Refs. on p. 151]

5 Nph = NO_2^- in the 5 position.) The order of conversion to high spin is 5 Meph < 5 Clph < 5 Nph, which is the order of the electron withdrawing strength. Substituents at other positions, as well as on bis complexes, gave similar correlations [6].

The conversion from high to low spin at modest pressure for the bis complexes is explicable in terms of increasing ligand field. We shall concern ourselves with the low spin to high spin conversion observed in the tris complexes and in many bis complexes at high pressure. The spin state is determined by the degree of back donation, which depends on the availability of the π^* orbitals. In Fig. 8.13 we show optical

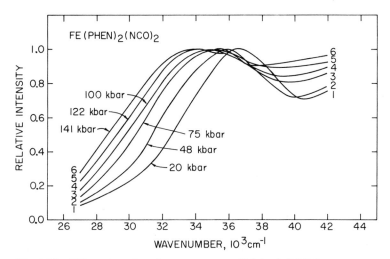

Fig. 8.13   UV spectra at various pressures – Fe(phen)$_2$(NCO)$_2$.

spectra in the region of the $\pi \rightarrow \pi^*$ peak for the bis isocyanate, which is typical for these compounds. There is a large red shift and significant peak broadening. As discussed in Chapter 6, the red shift is rather characteristic of $\pi \rightarrow \pi^*$ transitions. The shift is, however, only a modest fraction of the energy of the peak maximum. We postulate thermal occupation of the π^* level by π electrons by the process illustrated in Fig. 8.14. The reasons for a large difference between thermal and optical transition energies are discussed in detail in Chapter 3. Using Equation (3.22) of that chapter, and typical optical data for phenanthroline complexes, we obtain the results of Table 8.1. We assume $\omega' \simeq \omega$ although $\delta E_{1/2}$ is pressure-dependent. We observe that at low pressure the π orbital is stable by well over an electron volt, but above 50 kbar the sign of E_{th} changes, i.e., the π^* orbital is stabilized. [*Refs. on p. 151*]

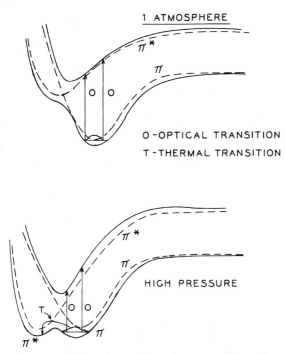

Fig. 8.14 Schematic configuration co-ordinate diagram: O, optical transition; T, thermal transition.

The analysis is, of course, very crude, but it is of interest that the sign change for E_{th} occurs near (just above) the pressure range where the low to high spin transformation originates. As the probability that the ligand electron has transferred from the π to the π^* orbital increases, the ligand π orbital becomes more available for back donation from the metal d_π electron. However, this transfer would tend to destabilize the d_π orbitals and decrease the ligand field.

One might expect the relocalization to be reflected in the high spin

Table 8.1 Thermal vs optical transitions – π-π^* transition for a phenanthroline complex

P, kbar	$h\nu_{max}$ eV	$\delta E_{1/2}$ eV	E_{th} eV
0	4.6	0.95	+1.35
50	4.45	1.05	+0.45
100	4.30	1.14	−0.40
150	4.20	1.20	−0.98

[*Refs. on p. 151*]

isomer shift. As discussed in Chapter 6, the normal behavior for ionic compounds is a decrease in the isomer shift with pressure due to spreading of the 3d orbitals. In Fig. 8.15 we see data for the bis isothiocyanate. There is an increase in the IS in the low pressure region, followed by a maximum. As we discussed in Chapter 6, the effect of ligand-metal interaction on the IS is complex. In this case the increase is consistent with the relocalization theory proposed above. The decrease at higher pressures may be associated with normal 3d orbital spreading, plus some back donation into the π orbitals as they become available.

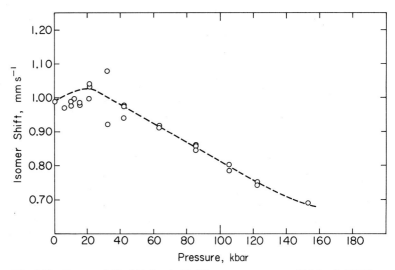

Fig. 8.15　Isomer shift of high spin Fe(II) versus pressure – $Fe(phen)_2(NCS)_2$.

It is clear that, for a series of these complexes at a given pressure, the amount of high spin should depend on the size of the ligand field resulting from the various interactions discussed above. Unfortunately, the direct optical measurement of Δ is not possible because the ligand field peaks are hidden by intense charge transfer transitions. However, we noted earlier that there was a good correlation between the low spin isomer shift and the spectrochemical series (the size of Δ). Thus, low isomer shifts correspond to large values of Δ, and should indicate relatively large amounts of low spin. In Fig. 8.16 we see such a correlation at 100 kbar for bis and tris phenanthroline and the bipyridil complexes studied by Fisher [5]. There is considerable consistency, especially for a series of bis compounds with different anion ligands. Fig. 8.17 shows a similar correlation from Bargeron's work [*Refs. on p. 151*]

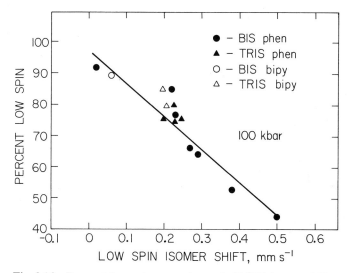

Fig. 8.16 Per cent low spin versus low spin Fe(III) isomer shift –
bis and tris phenanthroline complexes.

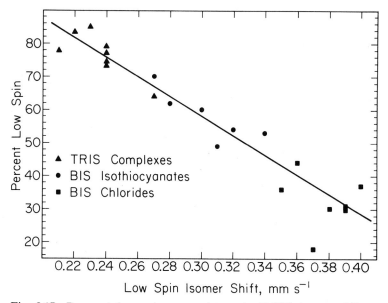

Fig. 8.17 Per cent low spin versus low spin Fe(III) isomer shift –
substituted phenanthroline complexes.

[*Refs. on p. 151*]

[6] for substituted phenanthroline. Again the consistency is quite satisfactory. Similar correlations hold at other pressures. As pointed out in Bargeron's paper, deviations from the correlation can be associated with steric effects.

In the visible region of the optical spectrum there exist absorption peaks which are generally assigned to metal to ligand charge transfer. In Fig. 8.18 we see typical spectra at various pressures for a tris phenanthroline complex, $Fe(phen)_3(N_3)_2 \cdot 6H_2O$. There is a single broad peak at 18 000 to 19 000 cm^{-1} which shifts to lower energy by 1500 to 2000 cm^{-1} in 140 kbar. Fig. 8.19 shows corresponding data for the bis complex, $Fe(phen_2(NCO)_2$. Here one observes two charge transfer peaks split by some 3000 cm^{-1}. Most bis phenanthroline complexes exhibit this splitting, but it is somewhat larger than usual for the cyanates, which makes the resolution easier. The splitting remains constant or decreases slightly with increasing pressure.

In the tris complexes the local symmetry is D_3 so that the metal d_π orbitals, of t_{2g} symmetry in a truly octahedral field, split to give a non-degenerate a_1 and a doubly degenerate level of e symmetry, with the former usually lower in energy. The splitting is usually a few hundred wave numbers. The bis complexes, which are in the cis conformation, exhibit C_{2v} symmetry which further splits the e level into non-degenerate states of b_1 and a_2 symmetry, but this splitting is very small.

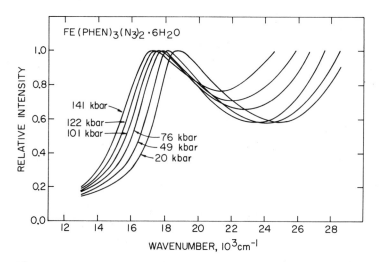

Fig. 8.18   Visible spectra at various pressures – $Fe(phen)_3(N_3)_2 \cdot 6H_2O$.

[*Refs. on p. 151*]

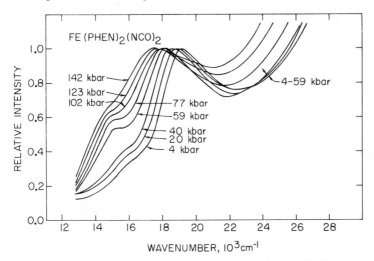

Fig. 8.19   Visible spectra at various pressures – Fe(phen)$_2$(NCO)$_2$.

Some interesting information can be obtained from the peak intensities. These are established by calculating the area under the peaks after subtracting out the tail of the π-π^* absorption. The intensity of the peak in the tris phenanthroline spectrum is relatively independent of pressure. For the sum of the two peaks in the bis phenanthroline spectrum there is about a 20% increase in the first 40 kbar, after which the area remains constant. The fraction in the low energy peak is near 10% at low pressure. By 40 kbar it has risen rapidly to about 20% and remains fairly constant at that fraction at higher pressures. The increase in total intensity in the first 40 kbar may be associated with the rapid high spin to low spin conversion in this region.

The most definite feature is the increase in relative intensity of the low energy charge transfer peak. Presumably the transfer is from the nearly degenerate a_2 and b_1 levels mentioned above to a ligand π^* orbital. For the low pressure high spin state there is a total of two electrons in these orbitals. For the low spin state there will be four electrons in these orbitals, and at 40 kbar a major fraction of sites has been converted to low spin. If there is no change in the relative transition probabilities for the two peaks, doubling the number of electrons would double the integrated intensity. Since only a portion of the sites are converted, this relationship must be regarded as approximate.

These low spin to high spin transitions then constitute a case where an electronic transition on the ligand (i.e., occupation of the ligand π^*

[*Refs. on p. 151*]

orbital by ligand electrons) changes the bonding characteristics of the ligand to the metal, and this modified bonding changes the field at the metal ion sufficiently to allow a spin rearrangement.

8.2.2 Ferrocyanides

The best known group of low spin ferrous compounds are the ferrocyanides. High pressure Mössbauer resonance studies have been made on the three heavy metal ferrocyanides $Zn_2Fe(CN)_6$, $Ni_2Fe(CN)_6$ and $Cu_2Fe(CN)_6$ as well as on $Na_4Fe(CN)_6$ and $K_4FE(CN)_6$. The former have a cubic structure [8, 9] with the local co-ordination $M - N \equiv C - Fe$. The latter two have a more complex arrangement with the cations arranged interstitially [10], but with the iron still co-ordinated to the carbon. In these compounds the iron is very strongly bound. The ligand fields are typically in the neighborhood of 30 to 35 000 cm^{-1} and the isomer shift is in the range -0.02 to -0.07 mm s^{-1} with respect to iron metal.

It would seem very unlikely that these compounds would undergo a spin change with pressure, and at room temperature no such event is observed in any of the compounds. At higher temperature and high pressure, one does observe significant amounts of high spin iron in some cases [11]. Fig. 8.20 shows the Mössbauer spectrum of $Cu_2 Fe(CN)_6$. The presence of a large amount of high spin ferrous ion at 189 kbar and 110 °C is unmistakable. The process is reversible but with considerable hysteresis. It is possible to recover samples with 15 to 20% high spin iron based on the Mössbauer spectrum and on susceptibility measurements on a very sensitive Faraday balance. Also, the infra-red spectrum of the recovered sample showed a shoulder at 2180 cm^{-1} (near the free cyanide value) as well as the peak at 2090 cm^{-1} typical for the ferrocyanides. The Fe-C stretching peak at 494 cm^{-1} had lost intensity and a new peak appeared at 467 cm^{-1}. This would be consistent with reduced Fe-C binding.

The conversion to high spin depends strongly on the cation as well as on pressure and temperature. Fig. 8.21 shows the conversion for $Ni_2Fe(CN)_6$ along isotherms at 110 °C and 147 °C. In Fig. 8.22 we compare conversions for the copper and nickel salts at 110 °C with that for $Zn_2Fe(CN)_6$ at 147 °C. (At 110 °C the zinc salt showed only traces of conversion at 200 kbar). The sodium and potassium salts exhibited only traces of conversion even at 200 kbar and 147 °C. The low spin to high spin transition in materials with such a large ligand field is puzzling, and the available explanations are not completely

[*Refs. on p. 151*]

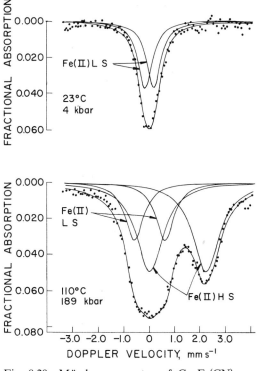

Fig. 8.20 Mössbauer spectra of $Cu_2Fe(CN)_6$ –
4 kbar (23 °C) and 189 kbar (110 °C).

satisfactory. Clearly the spin change must be involved with an interaction
of the cation electrons with the $(C \equiv N)^- \pi$ orbitals. There exist intense
ligand-metal charge transfer peaks in the ultra-violet (5 to 6 eV). They lie
too high in energy for convenient quantitative study in the high pressure
optical apparatus. However, they are clearly at lower energy for the heavy
metal cyanides than for the sodium and potassium salts. For the copper
salt, in particular, the charge transfer peak is low enough in energy to
mask almost all the crystal field peaks. It has also been established that
the charge transfer peaks apparently shift to lower energy and broaden
with increasing pressure. If we assume that these peaks represent cation
to ligand charge transfer, a very crude estimate would say that for
the copper salt there is a significant probability of occupation of the
cyanide π^* orbitals by copper d_π electrons above 100 kbar, even at
room temperature.

The change in isomer shift with pressure for the low spin Fe(II) ion
reflects the relative tendency for relocalization of the iron d_π electrons.

[*Refs. on p. 151*]

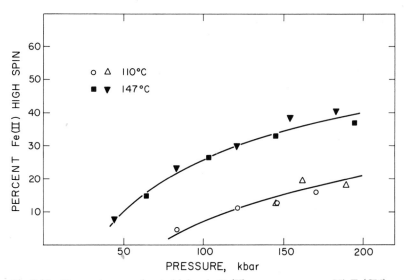

Fig. 8.21 Per cent conversion to high spin Fe(II) versus pressure – $Ni_2Fe(CN)_6$; 110 °C and 147 °C.

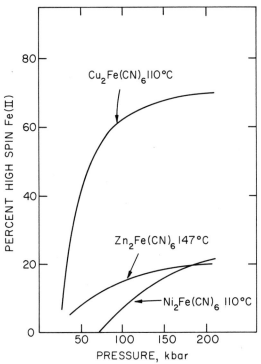

Fig. 8.22 Per cent conversion to high spin Fe(II) versus pressure – several ferrocyanides.

[*Refs. on p. 151*]

144

Fig. 8.23 exhibits the data for the five compounds at 298 K. The sodium and potassium salts show a decrease of the order normally observed in high spin compounds, associated with the general delocalization of the 3d electron cloud. The zinc compound shows slightly less decrease and the nickel compound distinctly less. The copper ferrocyanide shows an actual increase in isomer shift in the low pressure region, and a modest decrease at high pressure. Evidently the increased tendency for occupation of the 3d orbitals by the d_π electrons more than counterbalances the spreading of these orbitals. Both the relative location of the charge transfer peaks and the strong red shift of their low energy side would indicate that the decisive factor is the degree to which the ligand π^* orbitals are occupied by cation electrons.

The crystal field peaks of the heavy metal ferrocyanides in particular are not resolvable at high pressure. There have, however, been some measurements [12] of the relative value of B and Δ as a function of pressure for $K_3 Co(CN)_6$. B *increases* by perhaps 20% in 100 kbar, while Δ increases by 5 to 10% at low pressure and then levels off. This is in contrast to the behavior of typical high spin ions as discussed in Chapter 6, and is consistent with some relocalization of the $3d_\pi$ electrons.

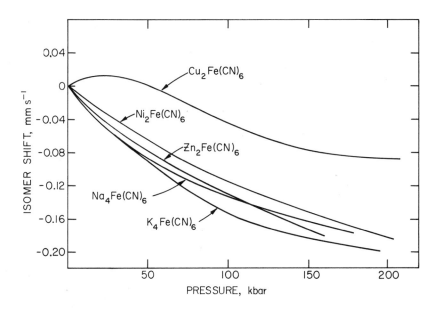

Fig. 8.23   Fe(II) low spin isomer shift versus pressure (relative to 1 atm) – several ferrocyanides.

[*Refs. on p. 151*]

8.2.3   *Phthalocyanine derivatives*

The study of iron in ferrous phthalocyanine and its derivatives offers an opportunity to measure the effect of pressure on the spin state of iron in a symmetry other than octahedral. In Fig. 6.22 of Chapter 6 we exhibit the structure of ferrous phthalocyanine, and its general similarity to the porphyrin derivatives whose structures are given in Fig. 6.25. In both cases iron is at the center of a planar molecule co-ordinated to four pyrrole nitrogens. The differences lie in the details of the outer structure of the molecule. The protoporphyrin IX is, of course, a more direct model for studying biological compounds of iron. However, it has the difficulty that in crystalline protoporphyrin the iron is ferric, whereas in hemoglobin it is ferrous. Since in phthalocyanine the iron is ferrous, there are some advantages associated with its study. In Chapter 10 we discuss the rather complex high pressure behavior of the iron porphyrins including both changes of oxidation state and of spin state. Here we follow the effect of pressure on the spin state of Fe(II) in phthalocyanine derivatives.

The local symmetry at the iron in ferrous phthalocyanine (FePc) is D_{4h}. When the fifth and sixth co-ordination sites are occupied by the same organic base, the local symmetry does not change. High pressure studies have been made on FePc and on derivatives with the axial sites occupied by pyridine, by picoline-3 (3-methyl pyridine), by picoline-4, and by piperidine (pentahydropyridine). The bonding of the ion in ferrous ion in FePc undoubtedly involves significant back donation of d_π electrons to the conjugated system of the ring. In the derivatives FePc(pyr)$_2$, FePc(pic-3)$_2$, and FePc(pic-4)$_2$, there is additional back donation to the conjugated electronic system of the axial ligands. Since piperidine has only saturated bonds, the backbonding of FePc(pip)$_2$ is entirely in the molecular plane.

The splittings of the 3d energy levels in D_{4h} symmetry are discussed in Chapter 2. Figure 8.24 shows three possible modes of occupation of these orbitals. The high spin state is in principle possible but is unlikely in compounds of true D_{4h} symmetry if there is even moderately strong interaction between the $d_{x^2-y^2}$ orbitals of the iron and the ligand nitrogen orbitals. (See, however, the discussion of hemin and hematin in Chapter 10.) The intermediate spin state may be the state of lowest energy here, as contrasted to the situation in O_h symmetry. It is, in fact, the ground state of FePc. With axial ligands, particularly if they bond strongly, the d_{z^2} orbitals are raised in energy sufficiently to give the low

[*Refs. on p. 151*]

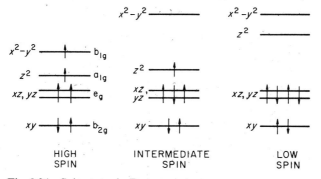

Fig. 8.24   Spin states in D_{4h} symmetry.

spin state as shown on the right. This is the ground state of the derivatives with axial organic bases. For pyridine and the picolines there is the additional factor that the d_π orbitals can backbond with the ligand π^* orbitals which further stabilizes the d_π orbitals. The effect of this additional backbonding is reflected in the energy of the charge transfer peaks. The stronger the backbonding, the higher the energy of the charge transfer transition. The picoline adducts studied have lower charge transfer energies that the pyridine adduct.

In the piperidine adduct the axial ligands have no empty π^* orbitals available for backbonding. The fact that it is low spin is due to the very strong σ bonding which raises the energy of the d_{z^2} orbital. One can see this from the magnitudes of the acid dissociation constant pK_A [13] which is 11.2 for the piperidine derivative compared with 5.25, 5.60, and 6.0 for the pyridine, 3-picoline, and 4-picoline adducts, respectively.

Fig. 8.25 exhibits the 1 atm Mössbauer spectrum of FePc. The isomer shift is 0.4 mm s^{-1} and the quadrupole splitting is 2.60 mm s^{-1}. With pressure there is a modest decrease in isomer shift and a considerable increase in quadrupole splitting but no new spin state appears [14]. (It is of interest to note that FePc is very sensitive to shear, and the sheared product has a significantly different Mössbauer spectrum. Yet, there was never as much as 10% of the shear product in any high pressure spectrum. This is evidence that shear is not a major factor in the observed pressure effects.)

Figure 8.26 exhibits the Mössbauer spectrum of FePc(pyr)$_2$ at 1 atm and at 103 kbar. At 1 atm the isomer shift is 0.27 mm s^{-1} and the quadrupole splitting is 2.0 mm s^{-1}. At 103 kbar there is an additional pair of peaks with larger isomer shifts and quadrupole splittings. The isomer shifts and quadrupole splittings for the two species

[*Refs. on p. 151*]

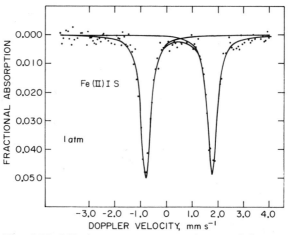

Fig. 8.25 Mössbauer spectrum at 1 atm of ferrous phthalocyanine.

are shown in Figs. 8.27 and 8.28, which also exhibit the results for FePc. In view of the close similarity between the parameters of the new phase and those of FePc, it seems quite reasonable to identify it as the intermediate spin ferrous state. It should be kept in mind, however, that the Mössbauer parameters are not always an unequivocal identification of the state, especially for low and intermediate spin species. The cause of the change is almost surely connected to reduction in backbonding by the same process as that described for phenanthroline earlier in this chapter. As can be seen in Fig. 8.29, the picoline adducts converted more to intermediate spin than the pyridine adduct. The methyl substitution in the picoline derivative reduces the ability to accept metal d_π electrons. This effect is reflected in the lower energy of the charge transfer peak [14]. At high pressure the conversion to intermediate spin stops. This appears to represent a balance between reduced backbonding and spreading of the 3d orbitals at high pressure. The latter effect favors a lower spin state. The delocalization of the 3d orbitals at high pressure is facilitated by the thermal transfer of ligand π electrons to the π^* orbital located on the periphery of the molecule, mentioned in Chapter 6.

The piperdine derivative showed the least conversion to intermediate spin, and at high pressure it actually reconverted to a higher percentage of low spin. The only backbonding is to the phthalocyanine nitrogens, so there is less decrease in total bonding strength with pressure. (The strong σ bond is not significantly affected.) At high pressures the spreading of the 3d orbitals is the controlling factor.

[*Refs. on p. 151*]

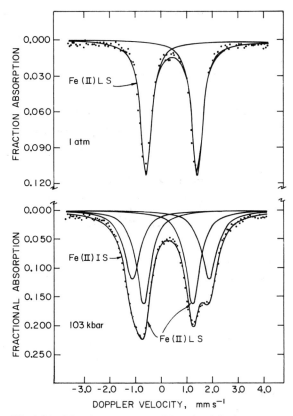

Fig. 8.26 Mössbauer spectra of FePc(pyridine)$_2$ at 1 atm and 103 kbar.

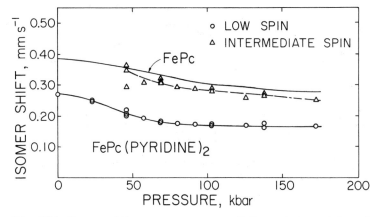

Fig. 8.27 Isomer shift versus pressure – FePc and two states of FePc(pyridine)$_2$.

[Refs. on p. 151]

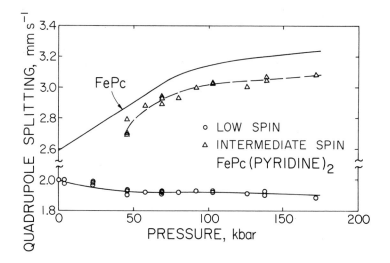

Fig. 8.28   Quadrupole splitting versus pressure – FePc and two states of FePc(pyridine)$_2$.

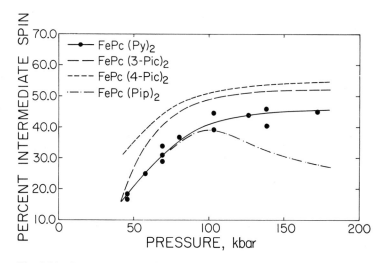

Fig. 8.29   Per cent conversion to intermediate spin versus pressure – phthalocyanine derivatives.

[*Refs. on p. 151*]

The studies in this chapter amply illustrate how changes in the electronic state of the ligand can affect the spin state of the metal co-ordinated to it.

References

1. J. S. GRIFFITH, *The Theory of Transition Metal Ions*, Cambridge University Press, London (1964).
2. R. W. VAUGHAN and H. G. DRICKAMER, *J. Chem. Phys.*, **47** 468 (1967).
3. C. B. BARGERON, M. AVINOR and DRICKAMER, *Inorganic Chem.*, **10** 1338 (1971).
4. N. E. ERICKSON in *Mössbauer Effect and Its Application in Chemistry*, edited by R. F. Gould, Amer. Chem. Soc., Washington (1967).
5. D. C. FISHER and H. G. DRICKAMER, *J. Chem. Phys.*, **54** 4825 (1971).
6. C. B. BARGERON and H. G. DRICKAMER, *J. Chem. Phys.*, **55** 3471 (1971).
7. P. DAY and N. SANDERS, *J. Chem. Soc.*, (A) 1530 (1967).
8. H. B. WEISER, W. O. MILLIGAN and J. B. BATES, *J. Phys. Chem.*, **46** 99 (1942).
9. A. K. VAN BEVER, *Rec. Trav. Chim.*, **57** 1259 (1938).
10. A. J. C. WILSON. *Struct. Rept.* **9** 209 (1942–44); **11** 421 (1947–48).
11. S. C. FUNG and H. G. DRICKAMER, *J. Chem. Phys.*, **51** 4353 (1969).
12. P. J. WANG, M. S. Thesis, University of Illinois (1971).
13. D. D. PERRIN, *Organic Complexing Reagents*, Chapter 4, p. 56, Interscience, New York (1964).
14. D. C. GRENOBLE and H. G. DRICKAMER, *J. Chem. Phys.*, **55** 1624 (1971).

CHAPTER NINE

The Reduction of Ferric Iron

A widely observed electronic transition discovered in iron compounds is the reduction from the ferric to the ferrous state. In a series of studies since 1967 such reduction has been observed in perhaps forty to fifty compounds including halides, cyanides, hydrates, salts of organic acids, and a variety of organometallic compounds [1]. Higher oxidation states, such as the ferrates, also reduce with pressure [2]. The mechanism involves the thermal transfer of an electron from a ligand non-bonding level to the metal d_π orbitals. As indicated in Chapter 6, the optical absorption peaks corresponding to ligand to metal charge transfer have maxima in the range 2 to 4 eV which shift to lower energy by as much as 0.2 to 0.4 eV in 150 kbar. The factors which permit a large energy difference between thermal and optical electron transfer are discussed in detail in Chapter 3. In this case the excited state consists of the ferrous ion and a collectively oxidized set of ligands. The presence of a 'hole' on the ligands decreases the probability that the Fe(II) site will exhibit typically ferrous behavior as regards metal-ligand vibrational frequency, etc. Much of the earlier data must be regarded as qualitative because of difficulties in perfecting experimental techniques. Furthermore, it is difficult to compare the earlier results due to the wide range of structural and bonding characteristics. These difficulties have been largely eliminated in recent studies on two series of acetylacetonate [3] and hydroxamate [4] derivatives. The examination of a related series of compounds with similar structural and bonding characteristics permits systematic variation of the electron donor properties of the ligand which can be correlated with the conversion. Refinements of technique have been used to minimize effects of sample

[Refs. on p. 170]   152 concentration, of shear, and other experimental artifacts discussed in Chapter 5. In addition, optical data complementing the basic Mössbauer data are available for these two series of compounds. For these reasons, we shall restrict our detailed discussion to these recent investigations.

The acetylacetonates have been extensively studied and are the subject of several reviews [5–7]. The structure of the twelve β-diketone ligands of the ferric acetylacetonate series studied has been illustrated already in Fig. 6.10 in the discussion of the isomer shift behavior. Typical Mössbauer spectra for the parent compound of the homologous series, $Fe(ACA)_3$, are shown at 23 °C and pressures of 1 atm and 41 and 170 kbar in Figs. 9.1 and 9.2. The atmospheric spectrum shows typical high spin Fe(III) isomer shift and quadrupole splitting. In many cases least squares fits at low pressure were particularly difficult to obtain due to low peak intensity and non-Lorentzian broadening arising from partially relaxed hyperfine

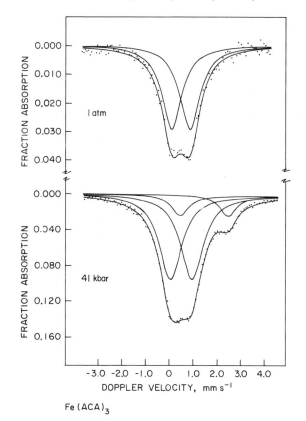

Fe (ACA)$_3$

Fig. 9.1 Mössbauer spectra of ferric acetylacetonate – 1 atm and 41 kbar.

[*Refs. on p. 170*]

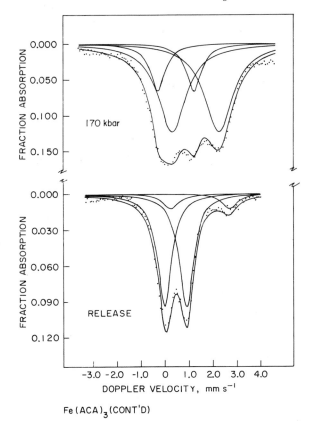

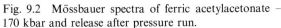

Fe(ACA)$_3$(CONT'D)

Fig. 9.2 Mössbauer spectra of ferric acetylacetonate –
170 kbar and release after pressure run.

spin-spin coupling [8]. The conversion to Fe(II) proceeds continuously
with pressure over the range above 30 kbar but does not go to completion.
After release of pressure and some effort to remove strain by powdering the
sample, the spectrum returns substantially to the ferric state (see lower part
of Fig. 9.2). The strain removal process is inefficient with such a small
sample; we feel that a complete release of mechanical strain would
probably give an entirely ferric spectrum. The observations that the
conversion occurs over a wide range of pressures and that there is
hysteresis upon release of pressure are associated with the fact that the
process is, in some degree, co-operative, with interaction between sites
being an important consideration. The analysis in Chapter 4 covers this
point in detail.

A useful means of presentation of the conversion data is in terms of
the equilibrium constant defined earlier as $K = C_{II}/C_{III} = AP^M$. Actual

[*Refs. on p. 170*]

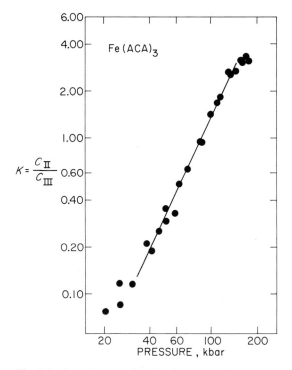

Fig. 9.3   Log K versus log P – ferric acetylacetonate.

data points for $Fe(ACA)_3$ are presented in this form in Fig. 9.3 to illustrate typical scatter. It is difficult to measure peak areas less than 10% of the total area under the envelope; here the conversion can be measured first at about 25 or 30 kbar. The linear behavior on the log-log plot expected from the empirical relation is obtained over a wide pressure range, with some tendency for deviation at the highest pressures.

Complementary evidence for the reduction process may be obtained from estimates of the areas under the respective charge transfer peaks in the optical spectra. There are two charge transfer bands in the β-diketonate derivatives (at 351 nm and 431 nm for $Fe(ACA)_3$), the assignment of which is still somewhat in dispute. There is general agreement that the higher energy peak corresponds to metal to ligand $(t_{2g} \rightarrow \pi_4^*)$ electron transfer with possibly some mixture of locally excited ligand character [9–11]. The lower energy peak has been variously assigned as $n \rightarrow e_g$ [9], $\pi_3 \rightarrow e_g$ [11], or $t_{2g} \rightarrow \pi_4^*$ [10]. It is clear, however, that neither of the peaks corresponds to the $\pi_3 \rightarrow t_{2g}$ transfer involved in the reduction process. For this reason, an analysis of thermal population

[*Refs. on p. 170*]

similar to the ones presented in Chapters 6 and 8 for $\pi - \pi^*$ transitions is not applicable. Nevertheless, the area under the charge transfer peak should reflect the concentration of ferric sites present.

The normalized areas of the lower energy charge transfer bands of three representative β-diketonate derivatives are shown in Figs. 9.4 to 9.6. The areas have been estimated by assuming Gaussian shapes and subtracting the large tail of the intense intraligand transition. The two charge transfer bands were treated independently; hence the 'background' subtracted involves both the $\pi \rightarrow \pi^*$ tail and a portion of the other charge transfer band. Thus the absolute values of the band width and intensities may be incorrect, but the trends in the area changes are considered a good first-order approximation. The data points and the solid curve refer to the normalized optical area; the dashed curve represents the decrease of ferric ion concentration as obtained from Mössbauer measurements. Two factors govern the pressure behavior of the charge transfer band areas. First, as indicated in Chapters 2 and 6, a general increase in the integrated charge transfer band intensity reflects an increase in transition moment with decreasing interatomic distance. This is expected on the basis of Mulliken's theoretical work on charge transfer spectra [12, 13]. On the other hand, a decrease in intensity would accompany the loss of ferric sites as they are converted to ferrous sites. A competition between the two effects is expected. Indeed, as can be seen from Figs. 9.4 to 9.6, the relative area under the charge transfer peaks decreases with increasing

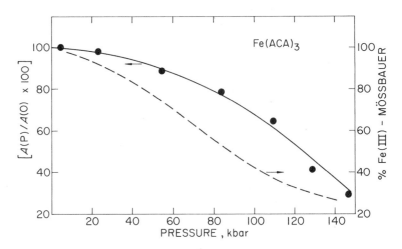

Fig. 9.4 Comparison of Fe(III) to Fe(II) conversion from optical and Mössbauer studies – Fe(ACA)$_3$.

[*Refs. on p. 170*]

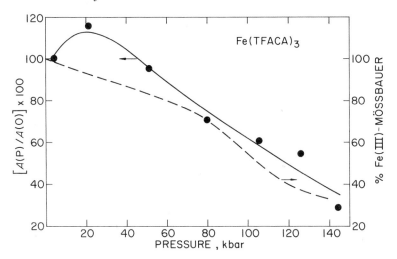

Fig. 9.5 Comparison of Fe(III) to Fe(II) conversion from optical and Mössbauer studies – Fe(TFACA)₃.

pressure, by an amount that closely parallels the conversion measured by Mössbauer resonance.

In analyzing the factors influencing the degree of conversion of Fe(III) to Fe(II) with pressure, it is desirable to have a measure of the tendency of the ligand to donate or withdraw electrons at high pressure. There are a number of possible measures of this tendency at 1 atm which, as we shall show, correlate well with the ferric isomer shift.

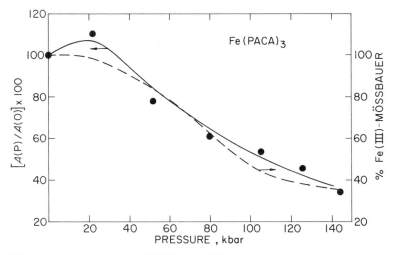

Fig. 9.6 Comparison of Fe(III) to Fe(II) conversion from optical and Mössbauer studies – Fe(PACA)₃.

[*Refs. on p. 170*]

One measure of the electronic character of the β-diketone is the acid dissociation constant associated with the enol form in the keto-enol equilibrium. Substituent effects are analyzed in terms of the tendency to increase or reduce the electron density around the oxygen atoms in the negatively charged anionic form. Electron donating groups will increase the oxygen electron density and destabilize the charge distribution, thus lowering the acidity and increasing the pK_D value. Electron withdrawing groups will have the opposite effect. Numerous measurements have been made of the acid dissociation constants [14–16].

A second semiquantitative relationship between chemical structure and electron donor-acceptor ability is given by electrophilic substitution constants. In general, such correlations have been applied mainly to aromatic systems where both inductive and resonance effects are operative. However, to the extent that the chelate ring is quasi-aromatic in nature, it is reasonable to attempt to use electrophilic substitution constants to describe the electronic properties of the metal chelate derivatives. The most common method of doing this is by means of the Hammett σ values, which are characteristic of the substituent added to the parent structure. Substituents with positive σ values are stronger electron acceptors than hydrogen; negative values indicate a weaker tendency for electron attraction than hydrogen. The σ values reflect a combination of inductive and resonance effects and are sensitive to the position of substitution. Brown and Okamoto [17] have shown that better correlations can be obtained for electrophilic reactions using slightly modified σ values. Both sets of substituent values work equally well for the compounds; the σ^+ set has been used here.

The third chemical parameter which can be related to the electronic behavior of the ligand derivatives is the appearance potential from electron impact mass spectrometry. The technique involves bombarding the substance in the gas phase with electrons and monitoring the ion current produced as the range of accelerating voltages is scanned. The appearance potential of interest in the metal β-diketonate systems corresponds to that voltage at which the singly ionized ML_3^+ species appears. Numerous examinations of the β-diketonate derivatives of the first transition series metals [18–20] have indicated that the appearance potentials depend predominantly on the ligand and only slightly on the metal. Thus data for the homologous series of Cu(II) chelates have been used because data for the Fe(III) chelates were incomplete.

Still another chemical parameter from which electronic information may be deduced is the half wave potential $E_{1/2}$ from polarography.

[*Refs. on p. 170*]

Electron donating groups on the chelate ring will tend to increase the basicity of the oxygen atoms and impart strong covalent character to the metal-oxygen bond. This will result in a large negative value of the half wave potential. Withdrawing groups lead to more ionic metal-oxygen bonds and a less stable chelate with a less negative half wave potential. As was the case for the appearance potentials, no literature data were available for the iron series derivatives. However, investigations of the copper series [21, 22] have been made and are used here.

Variations in the order of electron donor ability at 1 atm among the derivatives, as predicted by these different chemical correlations, do exist. However, three rough groupings of compounds are readily apparent. These are, in decreasing order of electron donor ability: [DPM(3), MACA(9), PACA(10), EACA(12)] > [ACA(1), DBM(2), BA(4)] > [TFACA(5), FTFA(6), TTFA(7), BTFA(8), NACA(11)] where the members of each general group have been arbitrarily arranged in order of their reference codes.

Figs. 9.7 and 9.8 show the correlation between atmospheric Fe(III) isomer shifts determined in this work, and the various atmospheric chemical measurements obtained from the literature. The isomer shift is given relative to bcc iron and the chemical parameters have been plotted so that movement to the right corresponds to an increase in electron donor tendency. The numbers refer to the reference codes given in Fig. 6.10. The same general correlation is observed in all cases, i.e., the smaller isomer shift, or greater s electron density at the iron nucleus, may be

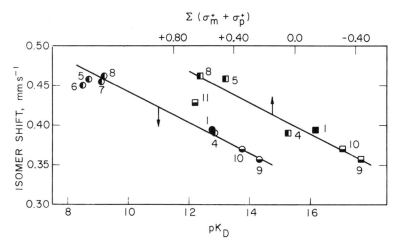

Fig. 9.7   Fe(III) isomer shift versus pK_D and Hammett σ.

[*Refs. on p. 170*]

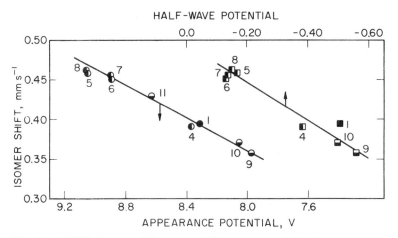

Fig. 9.8   Fe(III) isomer shift versus half wave potential and appearance potential.

associated with a greater tendency for donation of the ligand electrons to the metal. It is important to realize, however, that the absolute value of the isomer shift depends on several contributions, as has been pointed out by Erickson [23]. The isomer shift is affected by metal orbital expansion due to the reduction of effective nuclear charge associated with the overlap of the metal electron cloud with the negative ligand charge, and by 4s orbital augmentation, which constitute central field covalency. In addition, metal d_π backbonding to vacant ligand π^* orbitals or 3s shielding because of overlap of ligand electron density in the bond region, which constitute symmetry restricted covalency, are important factors to consider. Since these contributions may well exhibit different pressure behavior, the fact that the correlations of isomer shifts and measures of electron donor tendencies at atmospheric pressure are fairly consistent does not necessarily imply that such a correlation of absolute isomer shifts with degree of conversion will be as good at high pressures.

It is convenient to use here the classifications (A, B, and C) introduced in Chapter 6 to discuss the isomer shift. Smoothed conversion data for the three classes are shown in Figs. 9.9 to 9.11. Class A derivatives show continuously increasing conversion over the whole pressure range, with the linear behavior obtained over a substantial portion. Class B derivatives show tendencies to level off and become independent of pressure above 150 kbar. This levelling is accentuated further in Class C.

It appears from the log K-log P plots that there is a tendency for the conversions to converge at high pressures. Thus derivatives with large

[*Refs. on p. 170*]

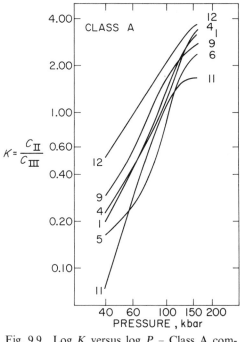

$$K = \frac{C_{II}}{C_{III}}$$

Fig. 9.9   Log K versus log P – Class A compounds.

conversions have small slopes and vice versa. This is demonstrated quite well in a plot of log A versus M in Fig. 9.12. All of the derivatives may be fitted with a straight line. Such a correlation is noteworthy because the values of A and M cover large portions of the ranges found between ionic and covalent systems. There is no important trend with respect to class distinctions but a distinction can be made on the basis of the relative order of electron donor ability presented earlier. Good electron donating groups tend to have large A and small M values; electron donors of intermediate strength have intermediate A and M parameters; and poor electron donors have small A and large M values.

To the extent that the absolute value of the ferric isomer shift is a measure of the tendency of the ligand to enhance thermal electron transfer, it should be possible to correlate the Fe(III) isomer shifts and the extent of conversion to Fe(II). Class A derivatives, which are characterized as generally poor π acceptors with a range of σ donor properties, are plotted in this form of correlation in Fig. 9.13. There is some scatter but it is clear that, for this group of compounds, a high isomer shift (low electron

[*Refs. on p. 170*]

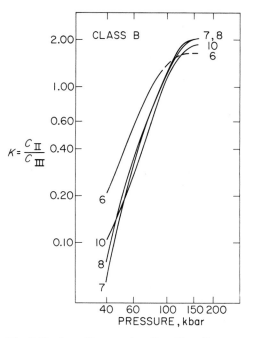

Fig. 9.10 Log K versus log P – Class B compounds.

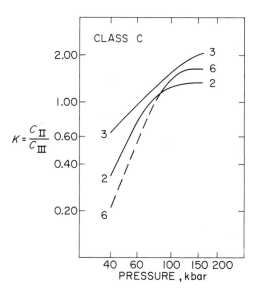

Fig. 9.11 Log K versus log P – Class C compounds.

[*Refs. on p. 170*]

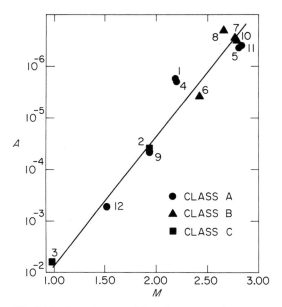

Fig. 9.12 Log A versus M – all compounds.

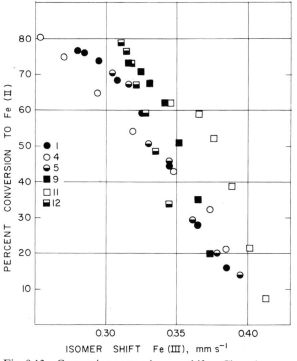

Fig. 9.13 Conversion versus isomer shift – Class A compounds.

[*Refs. on p. 170*]

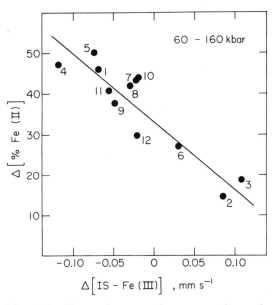

Fig. 9.14　Change in conversion versus change in isomer shift – (60–160 kbar).

density) may be associated with a small conversion and a low isomer shift (high electron density) with a high conversion.

The isomer shift depends in a complex way on both σ and π orbital overlaps, while the reduction of Fe(III) depends on the ability of the ligand π orbitals to transfer an electron to the metal d_π orbitals. In general, the energy difference between ligand and metal π orbitals decreases with increasing pressure and thus reduction proceeds. Among a series of related compounds such as the acetylacetonates it is to be expected that the relative increase in conversion with pressure will depend on the relative change of the electron donor ability as measured by the change in ferric isomer shift. In Fig. 9.14 we plot the increase of conversion with pressure between 60 and 160 kbar versus the change in isomer shift over the same range. The correlation holds quite well, i.e., those compounds which show a relatively large increase in donor ability, as measured by a large decrease in isomer shift, show a large increase in conversion, while those complexes which exhibit a relative decrease in donor ability (increase in isomer shift) show a relatively small increase in conversion.

The second series which allows for systematic variation of electronic properties is that of the hydroxamic acid derivatives with the general formula

[*Refs. on p. 170*]

where R can represent a variety of substituent groups. As in the acetylacetonates, the co-ordination to the ferric ion is through the oxygens after
removal of the acidic proton. There is renewed biochemical interest in
hydroxamic acids as a result of finding this characteristic group in natural
products from such organisms as aerobic microbial cells. The biological
function of the hydroxamates is to transfer iron through metabolic
channels and present it for incorporation into the porphyrins and other
iron-containing enzymes and proteins. The biological transfer of iron is
accomplished by a reduction of the ferric ion, which is bound tightly by the
hydroxamate groups, to the ferrous ion which is only weakly bound and
can be removed easily from the compound. The chemical and biological
aspects of hydroxamic acids have been examined by Neilands [24–27].

Three hydroxamate derivatives which are models for the biological
compounds and one biological hydroxamate, ferrichrome A, are included
in this series. The model compounds are tris (acetohydroxamato) iron (III),
tris (benzohydroxamato) iron (III), and tris (salicylhydroxamato) iron
(III) which will be referred to as $Fe(AHA)_3$, $Fe(BHA)_3$, and $Fe(SHA)_3$,
respectively. The ligands in these three hydroxamates are bidentate, as are
the β-diketones; in contrast, ferrichrome A is a hexadentate iron
compound.

The Mössbauer spectra of $Fe(SHA)_3$ at 4 kbar and 23 °C and at 138
kbar and 23, 110 and 135 °C are shown in Figs. 9.15 and 9.16. The low
pressure asymmetry of the ferric doublet indicates the presence of a spin-
spin relaxation effect similar to that observed in other materials. A larger
effect was observed for $Fe(BHA)_3$, but only a small relaxation effect was
found for $Fe(AHA)_3$. The relaxation times may be associated with the
distance between iron sites. As the pressure decreases the site to site
distance, the coupling between sites increases causing a shorter relaxation
time and an increase in the symmetry of the doublet. All of the derivatives
exhibited symmetric ferric peaks at high pressure. It is seen that at high
pressure the ferric iron reduces to ferrous iron, and with increasing
temperature the conversion increases significantly. This large increase of
reduction with increasing temperature occurs in most systems, indicating
that the process is quite endothermic. Fig. 9.17 shows the isothermal
yields in terms of equilibrium constant plots for $Fe(SHA)_3$. The yields for

[*Refs. on p. 170*]

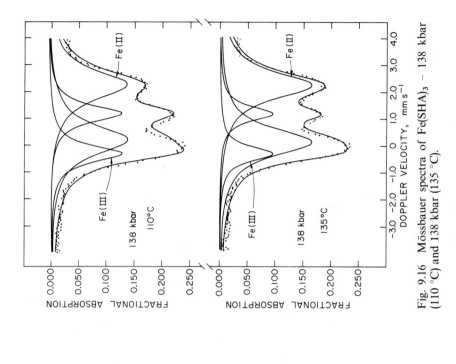

Fig. 9.16 Mössbauer spectra of Fe(SHA)₃ – 138 kbar (110 °C) and 138 kbar (135 °C).

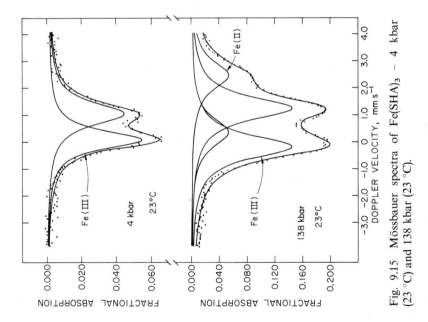

Fig. 9.15 Mössbauer spectra of Fe(SHA)₃ – 4 kbar (23 °C) and 138 kbar (23 °C).

[*Refs. on p. 170*]

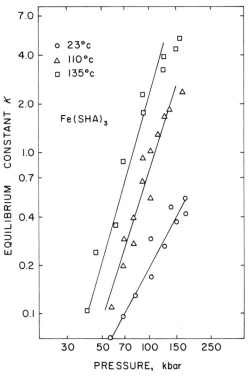

Fig. 9.17   Log K versus log P at 23 °C, 110 °C and 135 °C Fe(SHA)$_3$.

the three hydroxamates are compared at two different temperatures in Fig. 9.18.

The Mössbauer spectra of FA at 5 kbar and 138 kbar are shown in Fig. 9.19. Note that the lower pressure spectrum is broadened considerably more than that for Fe(SHA)$_3$ shown in Fig. 9.16. In FA this broadening has been attributed by Wickman *et al.* [28] to the presence of a magnetic hyperfine structure which is only partially relaxed at room temperature. At high pressures FA also reduces, but in greater yields than in the other three hydroxamates, as shown in Fig. 9.20 where the conversion of FA is compared to that of Fe(SHA)$_3$ at 23 °C.

High pressure optical data are available on the ligand to metal charge transfer bands for the hydroxamates. At 1 atm the peak maxima are at 23 250, 22 220, 21 950 and 22 220 cm^{-1} for Fe(AHA)$_3$, Fe(BHA)$_3$, Fe(SHA)$_3$, and FA, respectively. The peaks shift to lower energy, broaden, and lose intensity with increasing pressure. It is of considerable interest to compare conversions observed optically and by

[*Refs. on p. 170*]

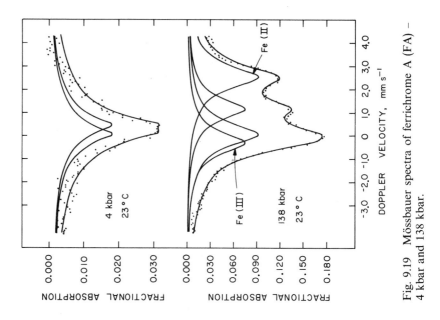

Fig. 9.19 Mössbauer spectra of ferrichrome A (FA) – 4 kbar and 138 kbar.

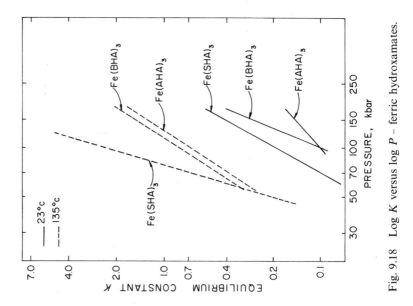

Fig. 9.18 Log K versus log P – ferric hydroxamates.

[Refs. on p. 170]

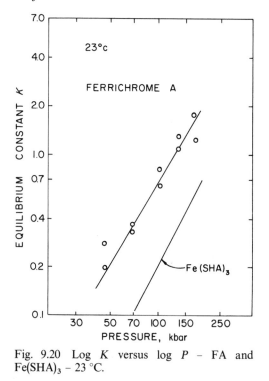

Fig. 9.20 Log K versus log P – FA and Fe(SHA)$_3$ – 23 °C.

Mössbauer resonance, and to apply the analysis of Chapter 3 to this series of compounds. Fig. 9.21 shows the semiquantitative agreement between the Mössbauer conversion measurements and optical charge transfer peak area estimations, similar to that observed in the acetylacetonates.

Although the assignment is not definite, if we assume the transition is ligand π to metal d_π, it is possible to use Equation (3.22) to calculate the thermal transition energy for the hydroxamates and ferrichrome A. Table 9.1 shows the results of such a calculation. It is seen that the

Table 9.1 Optical versus thermal transitions: ferric hydroxamates and ferrichrome A. For 10% reduction of Fe(III).

Compound	Pressure kbar	$h\nu_{max}$ eV	$\Delta E_{1/2}$ eV	E_{th} eV
AHA	125	2.80	0.90	−0.11
BHA	105	2.70	0.875	−0.06
SHA	70	2.54	0.84	−0.02
FA	37	2.65	0.835	+0.11

[*Refs. on p. 170*]

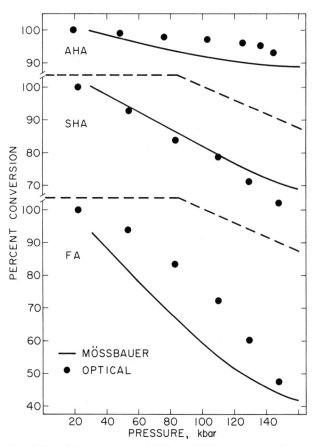

Fig. 9.21 Comparison of Fe(III) to Fe(II) conversion from optical and Mössbauer studies – Fe(AHA)$_3$, Fe(SHA)$_3$ and FA.

thermal energy approaches zero at pressures corresponding to 10% reduction.

In general, the studies on these two series of compounds demonstrate a high degree of consistency between optical and Mössbauer measurements, as well as the applicability of the analysis of Chapter 3 for the difference betwen optical and thermal processes.

References

1. H. G. DRICKAMER, V. C. BASTRON, D. C. FISHER and D. C. GRENOBLE, *J. Solid State Chem.*, **2** 94 (1970). This paper contains references to the work prior to 1970. Further discussion of reduction of Fe(III) occurs in Chapter 10.
2. V. N. PANYUSHKIN and H. G. DRICKAMER, *J. Chem. Phys.*, **51** 3305 (1969).

3. C. W. FRANK and H. G. DRICKAMER, *J. Chem. Phys.*, **56** 3551 (1972).
4. D. C. GRENOBLE and H. G. DRICKAMER, *Proc. Nat. Acad. Sci.*, **68** 549 (1961).
5. J. P. FACKLER, *Prog. Inorg. Chem.*, **7** 361 (1966).
6. J. P. COLLMAN, *Angew. Chem. Internat.*, **E4** 132 (1965).
7. D. W. THOMPSON, *Structure and Bonding*, **9** 27 (1971).
8. J. W. G. WIGNALL, *J. Chem. Phys.*, **44** 2462 (1966).
9. Y. MURAKAMI and K. NAKAMUR, *Bull. Chem. Soc. Japan*, **39** 901 (1966).
10. I. HANAZAKI, F. HANAZAKI and S. NAGAKURA, *J. Chem. Phys.*, **50** 265, 276 (1969).
11. R. L. LINTVEDT and L. K. KERNITSKY, *Inorg, Chem.*, **9** 491 (1970).
12. R. S. MULLIKEN, *J. Am. Chem. Soc.*, **74** 811 (1952).
13. R. S. MULLIKEN, *J. Phys., Chem.*, **56** 801 (1952).
14. L. G. VAN UITERT, C. G. HAAS, W. C. FERNELIUS and B. E. DOUGLAS, *J. Am. Chem. Soc.*, **75** 455 (1953).
15. L. G. VAN UITERT, W. C. FERNELIUS and B. E. DOUGLAS, *J. Am. Chem. Soc.*, **75** 457 (1953).
16. D. F. MARTIN and B. B. MARTIN, *Inorg. Chem.*, **1** 404 (1962).
17. H. C. BROWN and Y. OKAMOTO, *J. Am. Chem. Soc.*, **80** 4979 (1958).
18. G. M. BANCROFT, C. REICHERT, J. B. WESTMORE and H. D. GESSOR, *Inorg. Chem.*, **8** 474 (1969).
19. C. REICHERT, G. M. BANCROFT and J. B. WESTMORE, *Can. J. Chem.*, **48** 1362 (1970).
20. C. REICHERT and J. B. WESTMORE, *Can. J. Chem.*, **48** 3213 (1970).
21. H. F. HOLTZCLAW JR., A. H. CARLSON and J. P. COLLMAN, *J. Am. Chem. Soc.*, **78** 1838 (1956).
22. R. L. LINTVEDT, H. D. RUSSELL and H. F. HOLTZCLAW JR., *Inorg. Chem.*, **5** 1603 (1966).
23. N. E. ERICKSON in *The Mössbauer Effect and Its Application in Chemistry*, edited by R. F. Gould, Amer. Chem. Soc., Washington (1967).
24. J. B. NEILANDS, *Structure and Bonding*, **1** 59 (1966).
25. J. B. NEILANDS, *Bact. Rev.*, **21** 101 (1957).
26. T. E. EMERY and J. B. NEILANDS, *J. Am. Chem. Soc.*, **82** 3658 (1960).
27. J. A. GARIBALDI and J. B. NEILANDS, *J. Am. Chem. Soc.*, **77** 2429 (1959).
28. H. H. WICKMAN, M. P. KLEIN and D. A. SHIRLEY, *Phys. Rev.*, **152** 345 (1966).

CHAPTER TEN

Changes of Oxidation State and Spin State

In Chapter 8 we discussed changes in the spin state of ferrous iron with pressure. In Chapter 9 we treated the reduction of ferric to ferrous iron. We now discuss two types of systems where both processes are occurring. We again use Mössbauer resonance to identify the states present. The analyses of the Mössbauer spectra are considerably more complex for these systems so that the quantitative aspects are less precise than for the system discussed in the previous chapters, and, in some cases, even the identification of some states may be open to question. There are, however, some significant results even on a qualitative scale. First we investigate the state of iron in three iron porphyrin compounds [1], and secondly we analyze the mixed valence compound Prussian Blue (ferric ferrocyanide) in which a high spin ferric ion is coupled to a low spin ferrous ion through a cyanide bridge [2].

10.1 Iron porphyrins

The structure and biological function of the iron porphyrin molecule as a unit of hemoglobin is described in Chapter 6. (See Fig. 6.25.) For convenience we summarize the essential facts. The porphyrin molecule consists of four pyrrole rings bridged by methine groups. Various aliphatic groups are attached to the periphery – these are shown for the protoporphyrin IX used in this study in Fig. 6.25 of Chapter 6. The iron is essentially at the center of four nitrogens. Two types of derivatives will be considered. In imidazole protohemichrome (IMPH) two imidazole molecules are co-ordinated to the fifth and sixth sites. In hemin the fifth site is occupied by a Cl^- and in hematin by an OH^-. In both these cases the sixth site is unoccupied. In IMPH the iron remains essentially in the plane of the porphyrin ring. In hemin and hematin it is located at about 0.5 Å out of the plane, towards the Cl^- or OH^-. In all three molecules the iron is ferric iron. (This is a limitation on these molecules as biological prototypes because in hemoglobin one has four ferrous porphyrins.)

In Chapter 8 (Fig. 8.22) we discussed the possible spin state in D_{4h} symmetry. We pointed out that the high spin state was rather improbable; that, in the absence of axial ligands, ferrous phthalocyanine (FePc) is intermediate spin; and that the addition of axial ligands (organic bases) converted the system to low spin. In addition to the states mentioned above, there exists the possibility of a 'mixed spin' state, i.e., a very rapid equilibrium between two spin states which gives an effective average value between the two; thus a mixed spin-low spin may appear to be intermediate spin. FePc is true intermediate spin, but mixed spin states have been postulated for certain iron porphyrin compounds. It is not possible to distinguish between these states in the temperature range available in the high pressure apparatus. When, in connection with the iron porphyrins, we mention an intermediate spin state, one should understand the parenthetical expression 'or mixed spin.'

The ferric iron in IMPH is in the low spin state. The situation is quite analogous to the phthalocyanine with axial ligands. The combination of σ bonding and back donation raises the energy of the $d_{x^2-y^2}$ and d_{z^2} orbitals sufficiently so that they are not occupied. In hemin and hematin the iron is high spin. This is possible because the Fe (III) is 0.5 Å out of the plane of the porphyrin, which lowers the energy of the $d_{x^2-y^2}$ orbital sufficiently to give the high spin state.

Since both the Mössbauer spectra and the course of physical events we postulate are somewhat complex, we shall first state the overall conclusions we reach and then give the data and analysis we use to arrive at these conclusions. In all three compounds the ferric ion appears to reduce at high pressure. At 23 °C the hemin initiates measurable reduction below 20 kbar, hematin at slightly higher pressures, and IMPH near 50 kbar. This is reasonable as the charge transfer peaks are at 15 000 cm^{-1} and 15 500 cm^{-1} for hemin and hematin respectively, while for IMPH the charge transfer is not observed, so it must be buried under the more intense π-π^* transitions at higher energy. For all three compounds the material produced appears to be intermediate spin ferrous ion or mixed spin ferrous ion. The identification is made by comparison of Mössbauer parameters with those of FePc, or by comparing

[*Refs. on p. 184*]

differences of Mössbauer parameters in two spin states with their differences between spin states in phthalocyanine derivatives, as discussed in detail below. This is rather a tenuous means of identification but is the best available. One can give a reasonable justification for events leading to these states.

Let us first discuss IMPH. Fig. 10.1 shows spectra at 12 kbar and 103 kbar at 23 °C. The initial state is low spin ferric, as confirmed by susceptibility measurements. The initial isomer shift is 0.17 mm s^{-1} and the quadrupole splitting is 2.07 mm s^{-1}. With pressure there is a decrease in IS of ~0.2 mm s^{-1} in 170 kbar while the QS remains essentially constant, i.e., the behavior is normal. Above ~50 kbar a second pair of peaks appears with an IS of ~0.50 mm s^{-1} and a QS of ~2.15 mm s^{-1}. This evidence in itself is not sufficient to identify the new state. By the pressures at which one first indentifies a significant amount of

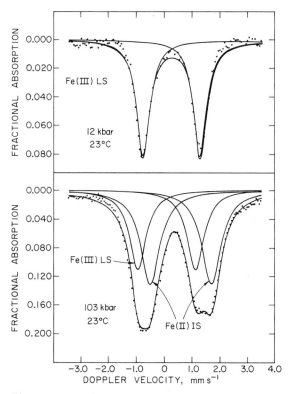

Fig. 10.1   Mössbauer spectra of imidazole protohemi-chrome – 12 kbar (23 °C) and 103 kbar (23 °C).

[*Refs. on p. 184*]

the new material, the evidence of Chapters 6, 8, and 9 indicates that there may be both thermal occupation of the charge transfer state by a ligand electron (giving reduction) and thermal occupation of the π^* orbital by a ligand π electron, reducing back donation, with possible change of spin state.

For most compounds, raising the temperature causes reduction at lower pressure. In Fig. 10.2 we exhibit the spectra of IMPH at 110 °C

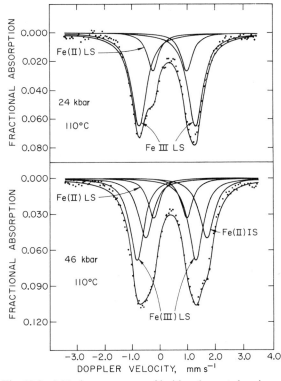

Fig. 10.2 Mössbauer spectra of imidazole protohemi-chrome – 24 kbar (110 °C) and 46 kbar (110 °C).

and at 24 and 46 kbar. At 24 kbar we observe a measurable amount of a material with an IS of 0.30 mm s^{-1} and a QS of 1.22 mm s^{-1}. Both the lower IS and smaller QS are reasonable for low spin ferrous ion. At this modest pressure there has been no significant reduction in the back donation and thus no change of spin state. At 46 kbar the spectrum can best be fitted by assuming three pairs of peaks: one like the pair discussed above, one like the original low spin Fe(III), and the third like the product produced above 50 kbar at 23 °C. As shown in

[*Refs. on p. 184*]

Fig. 10.3, at 103 kbar and 110 °C the low spin Fe (II) has disappeared. Yields of the two high pressure products as a function of temperature and pressure are shown in Fig. 10.4. A word of caution is advisable for these results and those for hemin and hematin discussed below. The quantitative yields are somewhat sensitive to the methods of sample crystallization and the previous sample history; however, the relative values shown here should be valid. The appearance of a product which is characteristic of low spin Fe (II) at low pressure and 110 °C and its transformation to a

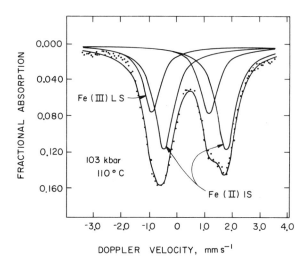

Fig. 10.3   Mössbauer spectrum of imidazole protohemi-chrome – 103 kbar (110 °C).

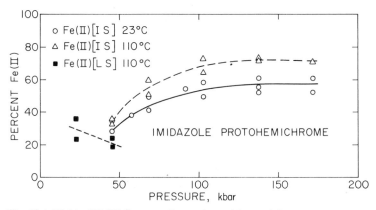

Fig. 10.4 Yield of Fe(II) (intermediate spin and low spin) versus pressure – imidazole protohemichrome.

[*Refs. on p. 184*]

high pressure product like the 23 °C high pressure product would indicate that the latter is intermediate spin (or mixed spin) Fe (II).

Further evidence is adduced from Fig. 10.5. Here we compare the difference in Mössbauer parameters (IS and QS) for the intermediate spin and low spin states of Fe (II) in FePc(pyr)$_2$ with the difference in the parameters for the two high pressure states formed in IMPH. The numbers are nearly identical. While neither of these items constitute proof,

Compound	Oxidation state	Spin state	Isomer shift	Quadrupole splitting
Phthalocyanines				
FePc	II	Inter.	0.35	2.90
FePc(Py)$_2$	II	Low	0.18	1.90
FePc(Py)$_2$	II	Inter.	0.34	2.85
Fe Pc (Py)$_2$		Δ (Inter.-Low)	0.16	0.95
Porphyrins				
Imidazole*	III	Low	0.05	2.05
Imidazole*	II	Low	0.28	1.25
Imidazole*	II	Inter.	0.45	2.25
Imidazole	II	Δ (Inter.-Low)	0.17	1.00
Hemin	III	High	0.32	1.25
Hemin	II	Inter.	0.27	3.00
Hermatin	III	High	0.33	1.40
Hematin	II	Inter.	0.27	2.80

* 110 °C.

Fig. 10.5 Isomer shifts and quadrupole splittings of various states of Fe(II) phthalocyanine derivatives and of iron porphyrin derivatives.

they are fairly strong evidence for the identification of the high pressure product as intermediate spin Fe (II). The reason for the increase in spin state is, of course, the reduction in backbonding due to thermal occupation of the π^* orbitals by ligand π electrons, as discussed in Chapters 6 and 8. There is no sign of a spin change for the ferric iron. This is probably associated with the fact that the ferric ion generally exhibits less back donation in a given situation then does the ferrous ion (e.g., ferri-cyanides exhibit about the same isomer shift as ferrocyanides although they have one less 3d electron to shield the 3s electrons).

The spectrum of hemin appears in Fig. 10.6. At 1 atm the peaks are equal in area but very different in shape. This is due to spin-spin coupling, as explained by Blume [3]. Since it is hard to establish the change in relative peak widths with pressure, there is an added problem in

[*Refs. on p. 184*]

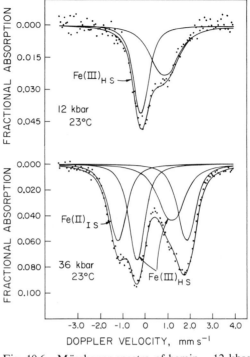

Fig. 10.6   Mössbauer spectra of hemin – 12 kbar
and 36 kbar.

obtaining quantitative fits at high pressure. As can be seen from the
lower parts of Fig. 10.6 and from Fig. 10.7, a new pair of peaks
appears in the spectrum and increases in size with increasing pressure.
As can be seen from Fig. 10.5, the Mössbauer parameters are very close
to those of FePc. This is taken as a reasonable identification of the
pressure-induced peaks as intermediate spin (or mixed spin) Fe (II). The
yield of this product as a function of pressure is shown in Fig. 10.8 for
hemin and in Fig. 10.9 for hematin. The slightly lower yields for
hematin may be due to the slightly higher energy of its charge transfer peak.
For hemin there is a slightly lower conversion at high temperature than
at 23 °C, while for hematin the yield is virtually independent of temperature. As indicated in Chapter 9, the amount of reduction is usually
considerably higher at high temperature. As discussed in Chapter 8, an
increase in temperature also favors a higher spin state. Since the
reduction here is accompanied by a lowering of the spin state, there is a
competition between the two effects of temperature.

The reduction in back donation at high pressure accounts for the

[*Refs. on p. 184*]

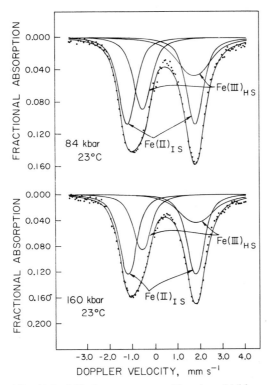

Fig. 10.7 Mössbauer spectra of hemin – 84 kbar and 160 kbar.

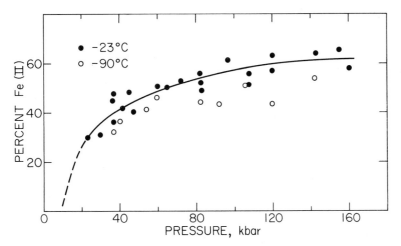

Fig. 10.8 Reduction of hemin versus pressure.

 [*Refs. on p. 184*]

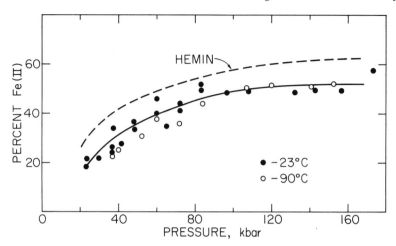

Fig. 10.9   Reduction of hematin versus pressure.

increase in spin state of the Fe (II) produced from IMPH. One must, however, account for the *reduction* in spin state accompanying the transformation of Fe (III) to Fe (II) in hemin and hematin. We recall that the ferric ion was high spin only because the Fe (III) ion was substantially out of the plane of the porphyrin ring. Zerner's [4] calculations for ferrous porphyrins indicate that if the metal were in the plane of the porphyrin ring it would be intermediate spin. If the iron were forced back towards the ring with pressure, the probability of intermediate spin would be enhanced. There is some evidence that this occurs. An analysis of the quadrupole splitting of Fe (II) in hemin by Moss [5] contains

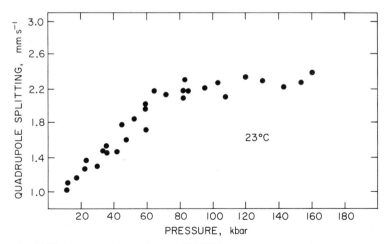

Fig. 10.10   Quadrupole splitting of Fe(III) versus pressure – hemin.

[*Refs. on p. 184*]

a term involving $1/z^3$ where z is the distance of the iron out of the plane. If the iron were forced back toward the plane by pressure, there would be a large increase in the Fe (III) quadrupole splitting, which is initially ~ 0.6 mm s^{-1}. In Fig, 10.10 is plotted the quadrupole splitting of Fe (III) in hemin versus pressure. By 100 kbar the splitting has increased by a factor of four. This is very reasonable evidence that the iron is indeed forced back towards the ring at high pressure. The problems involving peak shape are such that one cannot identify whether the ferric iron may be undergoing a reduction in multiplicity.

10.2 Prussian blue

Mixed valence compounds have been the subject of much discussion. A very thorough review is given by Robin and Day [6]. In these compounds a given element exists in two oxidation states with the possibility of electron transfer between them. Prussian Blue is ferric ferrocyanide. A high spin ferric ion is coupled to a low spin ferrous ion by a cyanide bridge. The coupling is Fe(III)-N $\equiv$ C-Fe(II), i.e., the first ferric ion couples to the nitrogen and the ferrous ion to the carbon. Robin [7] has shown that the electron has a 99% probability of being on the low spin ion co-ordinated to the carbon. It is possible to do an experiment tagging only the ferric ion with ^{57}Fe and following its spectrum as a function of pressure. Alternatively, one can tag only the ferrous ion. Both experiments have been performed [2]. Let us consider first the pressure effect on the ferric ion. In Fig. 10.11 one sees at 4 kbar the typical high spin ferric spectrum; at 145 kbar it is largely reduced to the high spin ferrous state. The yields are shown as a function of pressure in Fig. 10.12. The reduction at high pressure is surprisingly large. As we shall see below, there are two mechanisms for transfer of an electron to the Fe (II) ion. The experiments with the low spin ferrous ion tagged give more complex spectra. At 4 kbar one has the low isomer shift and neglible quadrupole splitting typical of low spin Fe (II). (See top of Fig. 10.13.) At moderate pressure and room temperature a pair of peaks appear with parameters typical of low spin *ferric* ion. At higher pressures, and particularly at high temperature, another pair of peaks appear which are clearly high spin *ferrous*. (See the bottom spectrum in Fig. 10.13). The yields are shown in Fig. 10.14. The low spin ferric ion appears at modest pressure, increases through a maximum near 120 kbar, and then decreases. The high spin ferrous ion appears in only modest quantity at any pressure at 23 °C, but the yield increases rapidly at higher temperature.

[*Refs. on p. 184*]

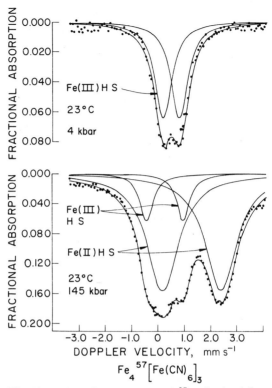

Fig. 10.11 Mössbauer spectra of $^{57}Fe_4[Fe(CN)_6]_3$, 4 kbar and 145 kbar.

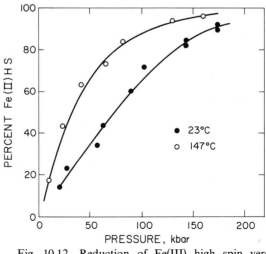

Fig. 10.12 Reduction of Fe(III) high spin versus pressure – $^{57}Fe_4[Fe(CN)_6]_3$.

[*Refs. on p. 184*]

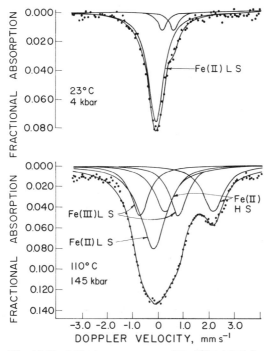

Fig. 10.13 Mössbauer spectra of $Fe_4[^{57}Fe(CN)_6]_3$, 4 kbar (23 °C) and 145 kbar (110 °C).

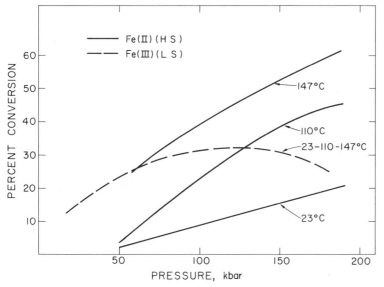

Fig. 10.14 Yields of Fe(III) low spin and Fe(II) high spin versus pressure and temperature – $Fe_4[^{57}Fe(CN)_6]_3$.

[Refs. on p. 184]

The processes involved are summarized in Fig. 10.15. One process is the reduction of high spin Fe (III) by the usual electron transfer from the ligand. The transfer of an electron from the low spin Fe (II) ion through

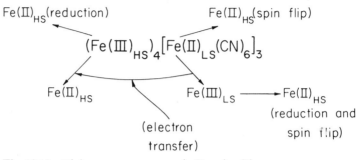

Fig. 10.15   High pressure processes in Prussian Blue.

the cyanide bridge to the high spin Fe (III) increases the yield of high spin Fe (II) at the site co-ordinated to the nitrogen and produces low spin Fe (III) at the site co-ordinated to the carbon. With increasing pressure and temperature this low spin Fe (III) reduces. Also at high pressure and temperature the back donation decreases, by the process discussed in Chapter 8, so that the low spin Fe (II) transforms to high spin Fe (II). The electron transfer across the cyanide bridge apparently depends little on temperature, in marked contrast to the other processes involved. The quantitative fitting of these complex spectra is difficult, but there seems little doubt about the process.

The events discussed in this chapter, even though they must be treated somewhat qualitatively, illustrate the richness and variety of the effects of pressure on electronic structure.

References

1. D. C. GRENOBLE, C. W. FRANK, C. B. BARGERON, and H. G. DRICKAMER, *J. Chem. Phys.*, **55** 1633 (1971).
2. S. C. FUNG and H. G. DRICKAMER, *J. Chem Phys.*, **51** 4353 (1969).
3. M. BLUME, *Phys. Rev. Letters*, **18** 305 (1967).
4. M. ZERNER, M. GOUTERMAN, and H. KOBAYASHI. *Theoret. Chim. Acta*, **6** 366 (1967).
5. T. H. MOSS, Ph. D. Thesis, Cornell University, New York (1965).
6. M. B. ROBIN and P. DAY, *Adv. Inorg. Chem., Radio Chem.*, **10** 247 (1967).
7. M. B. ROBIN, *Inorg. Chem.*, **1** 337 (1962).

CHAPTER ELEVEN

Reactions in Aromatic Molecules and Complexes

In Chapter 6 we saw that the energy of the optical transition (hv_{max}) between the highest occupied and lowest empty π orbitals ($\pi \rightarrow \pi^*$ transition) of aromatic molecules decreased rapidly with increasing pressure. Similarly, the optical absorptions associated with electron transfer between a donor and acceptor frequently decreased with increasing pressure. The theory of Chapter 3 indicated that a moderate decrease in this energy might be sufficient to permit thermal occupation of the excited state at high pressure, especially when configuration interaction and related effects were considered. In Chapters 6, 8, 9, and 10 we saw that for a variety of systems this thermal occupation was highly probable at high pressure, and that these electronic transitions led to new spin states and oxidation states for iron, particularly when complexed to aromatic or quasi-aromatic molecules.

In this chapter we discuss chemical reactions of aromatic molecules and of their electron donor-acceptor complexes with several electron acceptors. We discuss first the pure hydrocarbons, and then the complexes.

11.1 Hydrocarbons

Fig. 2.2 of Chapter 2 illustrates the electronic structure of the ground (A_1) state and of the first excited (^1L_a) state of one of the large aromatic hydrocarbons (anthracene). The ground state is non-polar, but there is a significant change in electron distribution upon excitation. In the solid state the larger polyacenes tend to be relatively insoluble and unreactive. For the series of polycenes starting with anthracene (3 rings)

185   [*Refs. on p. 208*]

through pentacene (5 rings) or hexacene (6 rings) there is little variation in arrangement of the molecules in the crystal lattice, and they have the same ground and first excited electronic states. As shown in Fig. 6.17 of Chapter 6, the v_{max} of the lowest energy optical transition decreases in energy at about the same rate with increasing pressure (or density) for anthracene, tetracene, and pentacene. However, pentacene (and hexacene) undergo irreversible chemical transformations at high pressure while anthracene and tetracene do not (at least below 500 kbar and in the absence of significant shear).

The major differences between, say, pentacene on the one hand and anthracene on the other are the values of v_{max} for the lowest $\pi \rightarrow \pi^*$ transition (17 000 and 27 000 cm^{-1} respectively) and the greater self-complexing tendency of the heavier hydrocarbons [1, 2].

We first review the experimental observations [3–5] and then discuss possible reaction mechanisms. The irreversible behavior is most apparent from electrical resistance measurements. The resistance of anthracene decreases with increasing pressure, but it always remains high, as does the activation energy for carrier production. In Fig. 11.1 we plot the

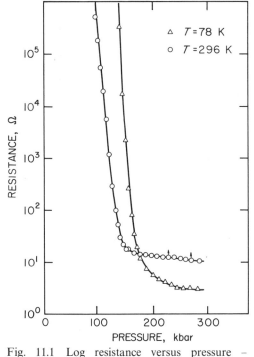

Fig. 11.1   Log resistance versus pressure – pentacene.

[*Refs. on p. 208*]

logarithm of the resistance of pentacene versus pressure at 296 and 78 K. Let us first consider the 78 K isotherm. The resistance drops by many orders of magnitude with increasing pressure. Below about 150 kbar the measurement is limited by the resistance of the cell. The curvature at high pressures is, in large part, a result of decreasing compressibility. At the same time the activation energy for carrier production, as measured from the temperature coefficient of resistance, decreases and goes to zero above 200 kbar. (See Fig. 11.2.) At higher pressures the resistance increases with temperature as for a metal, but the behavior [4] is more complex than for a simple metal. Undoubtedly there is a complex interaction of the Fermi surface with the Brillouin zone boundary. As long as the temperature is maintained below 180 to 200 K, this process is quite reversible.

Along the 296 K isotherm the resistance also decreases with pressure by orders of magnitude, but above about 200 kbar the resistance starts to drift upward with time. If the system is allowed to sit for 24 hours at pressure, it

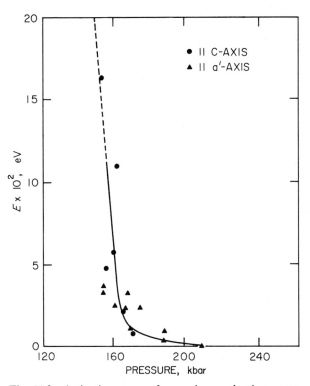

Fig. 11.2 Activation energy for carrier production versus pressure – pentacene.

[*Refs. on p. 208*]

may drift upward by an order of magnitude or more. The irreversiblity is illustrated more clearly in Fig. 11.3. Pentacene is compressed at 78 K to above 300 kbar; then the system is warmed (dotted vertical line). If one stays below 200 K the resistance increase is reversible. If the system is allowed to come to room temperature (open circle on dotted line), the process is no longer reversible, and upon cooling the resistance *increases*. On cooling from 296 to 78 K it increases by a factor of six or seven. Further compression decreases the resistance, but it now remains a semi-conductor at all pressures.

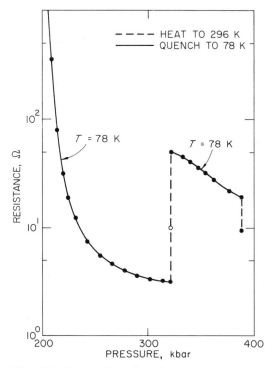

Fig. 11.3   Irreversible effects in pentacene.

The material can be recovered in milligram quantities and some rough characterization is possible. Ordinary pentacene is a bluish black crystalline material which sublimes easily at 110 °C in a vacuum. The recovered product is reddish brown and amorphous. Heating to several hundred degrees centigrade leaves it apparently unchanged. The electronic spectra of pentacene and of the high pressure product appear in Fig. 11.4. The pentacene is characterized by the low lying excited states near 600 to 700 nm ($\sim 17\,000$ cm^{-1}). These correspond to the $A_1 \rightarrow {}^1L_a$ excitation of

[*Refs. on p. 208*]

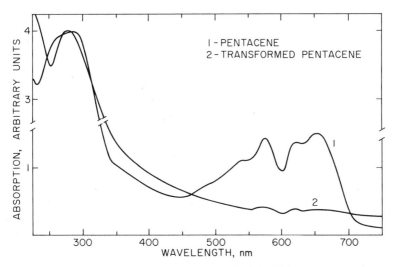

Fig. 11.4   Electronic spectra of pentacene and of the high pressure product.

Fig. 2.2(b) and involve electrons conjugated around the entire periphery. These have disappeared in the high pressure product, although there are still absorptions at higher energy which correspond roughly to those observed in benzene and possibly naphthalene.

The infra-red spectra of pentacene and of the high pressure product appear in Fig. 11.5. There are a number of differences. In the first place, the peaks in the spectrum of the high pressure product are considerably broadened. This is characteristic of polymers. In the high energy end of

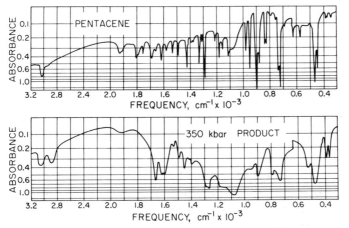

Fig. 11.5   Infra-red spectra of pentacene and of the high pressure product.

[*Refs. on p. 208*]

the spectrum there is a difference of particular interest. In pure pentacene the C—H stretching peak at 3030 cm^{-1} is characteristic of aromatic molecules. In the high pressure product there is a second peak of essentially equal intensity at 2900 cm^{-1}, the position characteristic of saturated C—H stretching vibrations. There are also broad peaks at 1670, 1600, and 1450, 1275, 1175, 1075, 800, 515, 395 and 370 cm^{-1} which do not appear in the spectrum of pentacene.

The high pressure product is essentially totally insoluble in all solvents so that its characterization is difficult. An observation of interest and some assistance is that a number of the peaks in the high pressure product appear also in pentacene peroxide produced photochemically (e.g. those at 3030, 2905, 1670, 1495 and 1450 cm^{-1}). There is, of course, no oxygen present in the high pressure cell; besides, the relative intensity of the 2900 cm^{-1} in the high pressure material is much greater. The similarities may be somewhat indicative of the bonding in the high pressure product.

Some speculations concerning the product are possible. Fig. 11.6 shows the molecular intersections with the *ab* plane for pentacene. Fig. 11.7 shows possible forms of polymer which could be formed by compression with a minimum of molecular displacement or rotation. It would be very

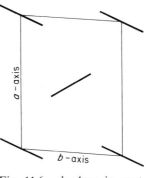

Fig. 11.6 *ab* plane in pentacene crystal. Scale 1 cm = 1 Å.

interesting to resolve the structure of the product. From the standpoint of the thesis of this monograph it is even more important to analyze the possible mechanism of reaction. The differences between pentacene, which reacts, and anthracene, which does not are: (a) the lower energy of the $\pi \to \pi^*$ optical absorption peak in pentacene (2.1 versus 3.3 eV) and (b) the greater self-complexing tendency of pentacene. These two effects are, of

[*Refs. on p. 208*]

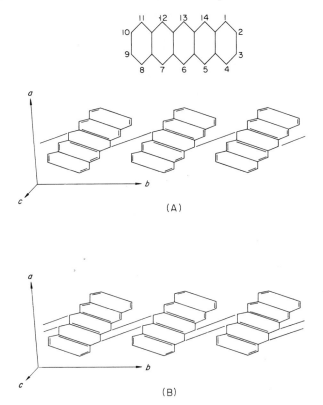

Fig. 11.7   Possible high pressure polymers of pentacene.

course, not unrelated. The $\pi \rightarrow \pi^*$ transition shifts 0.6 eV in 60 kbar and must shift at least 1 eV in 200 kbar, perhaps more. At the time these experiments were done it was possible to make only very crude estimation of half widths in the high pressure optical cell, but the peaks are broadening with increasing pressure. A half width of 0.5 eV at 200 kbar would easily provide for significant thermal occupation at that pressure, even without considerations of configuration interaction, etc. The charge transfer probability would also be enhanced by the relative lowering of the π^* orbital and by the broadening of bands of allowed energy. Eximers of the type $Ar : Ar^*$ and dimers of the form Ar_2^+ are known to form photochemically for many aromatic molecules. In any case, the process of occupying a new ground state puts the molecule into a reactive configuration. In anthracene the π^* orbital lies sufficiently high at all pressures so that the probability of significant thermal electronic excitation by either mechanism is negligible.

[Refs. on p. 208]

11.2   Reactions in electron donor-acceptor complexes

As discussed in the previous section, aromatic hydrocarbons may react at high pressure if there exists an excited state at sufficiently low energy so that it can be occupied thermally. The excited state may be a π^* orbital of the molecule in question, or a similar state of an adjacent molecule. In the latter case, one speaks of 'self-complexing.' In Chapters 2 and 6 we discussed the properties of electron donor-acceptor complexes and showed that the optical absorption associated with the electron transfer frequently lies at energies less than 3 eV and decreases with increasing pressure. It is not, therefore, surprising that at high pressure there should frequently be significant thermal occupation of the electron acceptor orbitals by a donor electron. In fact, electrical resistance measurements have shown that there are irreversible effects in a wide variety of such complexes at high pressure [5–7]. Typical electron donors involved include aromatic hydrocarbons and aromatic amines, while acceptors include iodine, tetrocyanoethylene, chloranil, and bromanil. Typically the reactivity is detected as an irreversible rise in resistance with time at high pressure. The most spectacular of such effects are in complexes involving tetracyanoquinodimethane (TCNQ) as the anion. These complexes typically have very low resistances at atmospheric pressure. In Fig. 11.8 we exhibit the resistance-pressure curve for the triethylammonium complex. Above 80 to 100 kbar the resistance tends to drift upward with time; this tendency accelerates with increasing pressure. By 400 kbar the resistance has increased by a factor of 10^3 or more. Below about 180 K the complex exhibits only a monotonic reversible decrease in resistance with pressure and no reactivity. This is a common feature of the electrical (and chemical) behavior of both the pure hydrocarbons and the complexes.

The general mechanism of reaction is reasonably evident. With pressure there is a decrease in energy of the excited state until it can be thermally occupied by a donor electron. This puts either the donor or acceptor, or both, into a reactive configuration so that reaction proceeds, limited, of course, by the geometrical constraints of the lattice.

Unfortunately, there has so far been only very modest progress in identifying the nature of the products, in part because of the small samples involved, and in part because insoluble polymers are frequently obtained. We discuss here only two sets of complexes. We describe briefly the behavior of complexes of hydrocarbons with tetracyanoethylene (TCNE) and, more extensively, hydrocarbon-I_2 complexes, where significant progress in product identification has been possible.

[*Refs. on p. 208*]

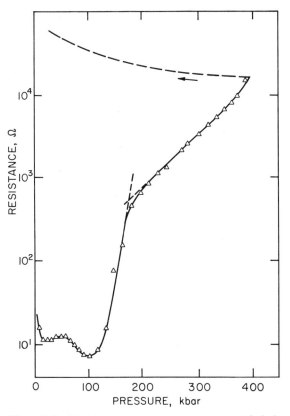

Fig. 11.8 Resistance versus pressure – triethyl-ammonium TCNQ.

11.3 Hydrocarbon-tetracyanoethylene complexes

In Fig. 11.9 we exhibit the resistance-pressure curve for the 1 : 1 complex of TCNE with the aromatic hydrocarbon perylene $(C_{20}H_{12})$. The irreversible behavior is clearly exhibited above 100 kbar, but actually initiates as low as 70 kbar. The product has been recovered in milligram quantities. In Fig. 11.10 the infra-red spectrum of the complex is compared with that of the high pressure product. The most obvious difference is the very broad absorption at energies below ~ 1700 cm^{-1} in the high pressure product compared with the sharp lines of the complex. Clearly, a rather complicated polymer has been formed. The complex exhibits a typical aromatic C—H stretching vibration at 3050 cm^{-1}, while the product has also measurable absorption near 2910 cm^{-1}, which

[*Refs. on p. 208*]

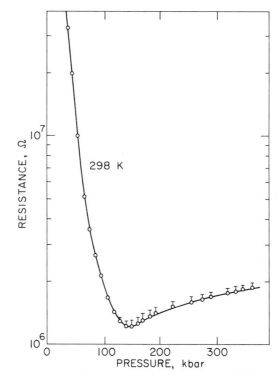

Fig. 11.9 Resistance versus pressure – TCNE-perylene.

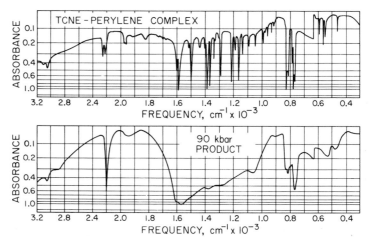

Fig. 11.10 Infra-red spectra of TCNE-perylene and of the high pressure product.

[*Refs. on p. 208*]   194

indicates the presence of some saturated C—H bonds. In the complex there are a pair of peaks at 2260 and 2220 cm^{-1} assigned to the C $\equiv$ N stretching vibration. The high pressure product exhibits only a single intense line at 2210 cm^{-1}. The fact that the TCNE is not separable from the product indicates that both donor and acceptor are involved in the reaction. Complexes of TCNE and naphthalene yielded a product which had a spectrum very much like that of the TCNE-perylene product.

Anthracene forms a Diels-Alder product with TCNE at 1 atm, with a very different spectrum and crystal structure [8] from those of the electron donor-acceptor complexes. The product formed at 225 kbar from the adduct was essentially identical with that shown in the lower half of Fig. 11.10. Evidently, with high pressure the adduct is destroyed and a polymer is formed.

It is possible to follow the course of the reaction for TCNE-perylene by observing the charge transfer peak, which lies near 11 000 cm^{-1} ($\sim$1.4 eV) at 1 atm, in the high pressure optical cell. Fig. 11.11 exhibits

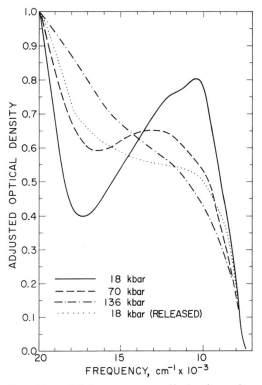

Fig. 11.11 High pressure optical absorption measurements, TCNE-perylene.

[*Refs. on p. 208*]

spectra at several pressures. The most obvious feature is the fading of intensity of the peak, starting at about 70 kbar. The spectrum run after release of pressure shows that the peak does not recover intensity. This is, of course, consistent with the irreversible nature of the chemical changes as discussed above. There is relatively little red shift of the peak with pressure, but measurable broadening. Calculations using Equation (3.22) of Chapter 3 indicate that at atmospheric pressure the normal ground state is thermally stable by only 0.2 to 0.3 eV. and that the broadening under pressure is sufficient to permit thermal transfer of electrons by 70 kbar.

11.4   Hydrocarbon-iodine complexes

The complexes of iodine with pyrene ($C_{16}H_{10}$) and perylene ($C_{20}H_{12}$) provide the most detailed information as yet available about the high pressure reactivity of CT complexes [9]. The complexes studied were the 2 perylene-$3I_2$ complex and pyrene-$2I_2$ complex. The former is estimated by Uchida and Akamatu [10] to have the structure shown in Fig. 11.12.

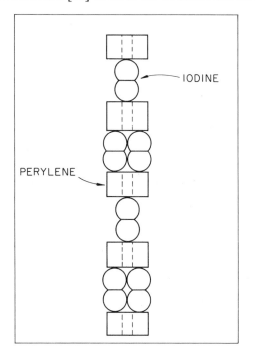

Fig. 11.12   Model of probable structure of the
2 perylene-$3I_2$ complex.

[*Refs. on p. 208*]

They base their estimate in part on their powder X-ray patterns and in part by analogy to the careful work of Hassel and Rømming [11] on the benzene-bromine complex. Presumably the pyrene-$2I_2$ complex has a similar layered structure. With care the stoichiometry of these complexes can be maintained within 0.5% of the ideal.

The pure hydrocarbons do not react at the highest available pressure. Figs. 11.13 and 11.14 exhibit the resistance-pressure curves for the complexes. The irreversible behavior is evident. (The vertical lines indicate the resistance drift in one minute's time.) In order to accumulate enough sample for analysis, the larger volume high pressure cell was used [5], and 25 to 30 consecutive 24 hour runs were made at pressures in the range 165–200 kbar. The iodine was removed by sublimation and came off quantitatively. It is important to note that although the iodine is necessary for the reaction, it doesn't enter into the final products. The recovered product was separated into portions soluble and insoluble in

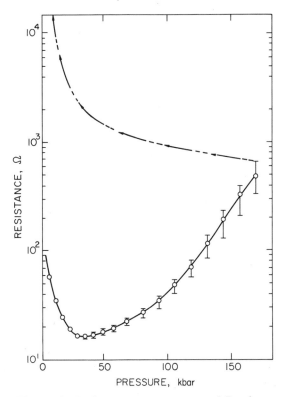

Fig. 11.13 Resistance versus pressure – 3 Perylene-$2I_2$ complex.

[*Refs. on p. 208*]

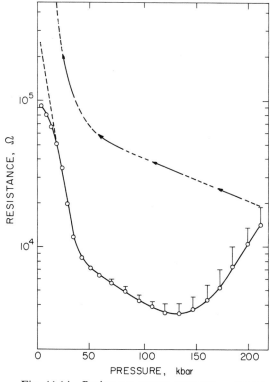

Fig. 11.14   Resistance versus pressure – pyrene-
$2I_2$ complex.

benzene. Since the insoluble portion could not be characterized we do
not discuss it further. The soluble material was fractionated by
chromatography into monomer and a series of polymer fractions. Two
fractions from the perylene and two from the pyrene product were
sufficiently large to permit a rather complete characterization. It is
important to keep in mind that these materials were not pure compounds,
but groups of closely related materials. Further fractionation would leave
insufficient materials for characterization. These were quite satisfactory
for establishing the *types* of products made. A combination of vapor
phase osmometry and mass spectroscopy indicated that both perylene
fractions were dimers, while both pyrene cuts were tetramers. Each cut
was characterized by ultra-violet and visible, infra-red and n.m.r. spectroscopy; some information was also gathered from the mass spectra. The
optical techniques were rather standard. The n.m.r. spectra were obtained
on a Varian HR-220 spectrometer using the technique of computer
averaging of transients (CAT). It was necessary to use 80 to 150 scans.

[*Refs. on p. 208*]

Three to five different runs reproducing the distribution of hydrogens were made on each sample. The general characteristics of the products as established from the various techniques appear in Table 11.1. We discuss here in some detail, and give a structure for, one dimer (A) of perylene and one tetramer (A) of pyrene.

Table 11.1 Structural features of products

	Pyrene tetramers		Perylene dimers	
	A	B	A	B
Number of protons	40	40	24	24
Total paraffinic protons	31	30	19	17
(a) normal ring protons	14	13	7	7
(b) 3-membered ring protons at $\delta = 1.1$	5	4	4	4
(c) protons shifted downfield	12	13	8	6
Total olefinic protons	3	5	2	4
(a) at normal position	0	1	1	1
(b) shifted upfield by benzene ring	3	4	1	3
No. of aromatic protons	6	5	3	3
3-membered rings	6	5	5	4
Total aromatic rings	2	2	2	2
aromatic rings with protons	2	2	1	1
Naphthalene chromophore	0	0	1	1
Total olefin bonds	8	9	4	5
olefin bonds with protons	2	3	2	3
New bonds′	18	17	11	10

The n.m.r., infra-red and electronic spectra for perylene dimer A appear in Figs. 11.15 to 11.17. Most of the structural information is obtained from the n.m.r. spectrum. Fig. 11.15 compares dimer A with pure perylene. The details of analysis of such a spectrum are given in a number of texts, e.g. [12, 13] and will not be reviewed here. The original paper [9] presents a complete analysis. The horizontal scale measures the shift downfield in parts per million (ppm) of the various proton resonances relative to the line of tetramethylsilane (TMS). The aromatic hydrogens show the largest shift, then the olefins, and then the paraffinic hydrogens in various configurations. The especially interesting features are the relatively few aromatic protons, and the large number (five) of three-membered rings. The infra-red spectrum (Fig. 11.16) is consistent in that the absorption in the region 2900 to 3000 cm^{-1} indicates paraffinic C-H stretching vibrations and the broad peak near 1100 cm^{-1} is due

[*Refs. on p. 208*]

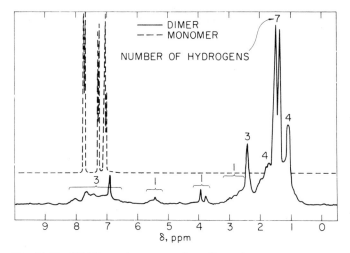

Fig. 11.15   NMR spectrum – perylene dimer A.

to ring bending on three, five, six and seven-membered rings. The visible and ultra-violet spectrum appears in Fig. 11.17. The most surprising feature is the low energy absorption at 430 to 460 nm, in view of the greatly reduced aromatic character of the product. Apparently the three-membered rings contain sp_2 hybrids plus p-like orbitals which have appropriate symmetry to conjugate with the π electrons of the olefins to give a relatively large conjugated chain.

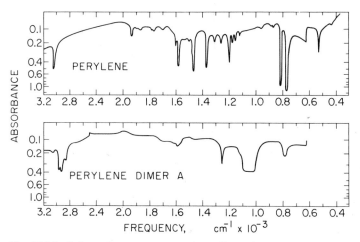

Fig. 11.16   Infra-red spectrum – perylene dimer A.

[*Refs. on p. 208*]

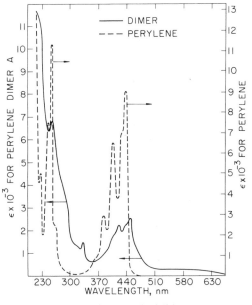

Fig. 11.17 Ultra-violet and visible spectrum –
perylene dimer A.

A model which reproduces the features of dimer A as listed in Table
11.1 appears in Figs. 11.18 and 11.19(a) and 11.19(b). Fig. 11.18 is a
stereoscopic illustration, i.e., if it is viewed through stereoscopic glasses
a three-dimensional representation of the molecule is observed.
Fig. 11.19(a) is a projection from about the same angle as Fig. 11.18,
while Fig. 11.19(b) gives a top view.

The model consists of two skew monomer layers. The molecule has
five three-membered rings, although the n.m.r. spectrum, Fig. 11.15,
suggests a minimum of four three-membered rings. The predicted
spectrum of the model, however, agrees with the actual spectrum since
one of the protons isolated on a three-membered ring is expected to be
shifted downfield by van der Waals interaction with a nearby carbon.
Protons which are on both three and five-membered rings, as well as
the shifted proton on one of the three-membered rings, would appear
near 1.5 ppm. There are ten of these protons. Five of these are shifted
downfield by nearby π bonds, but two other protons appear near 1.5 ppm
due to anisotropies of a σ bond.

One olefinic proton, most easily seen in Fig. 11.19(b), is positioned
approximately over the center of a benzene ring. This proton could well
appear slightly above 4 ppm. The other olefinic proton is in the same

[*Refs. on p. 208*]

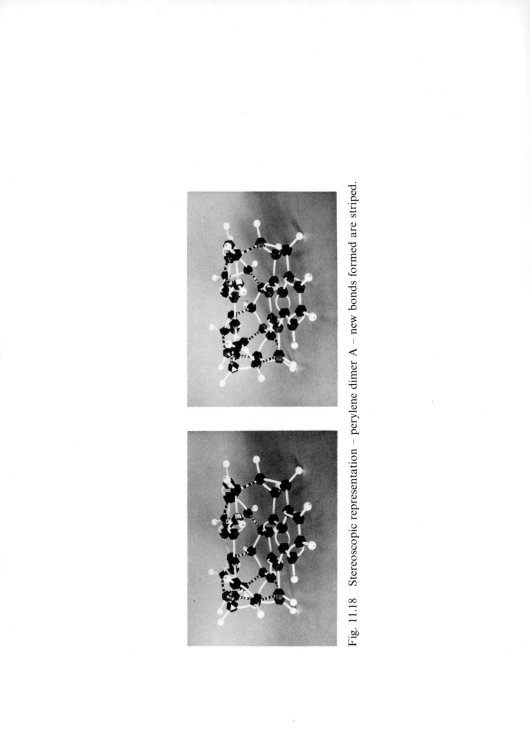

Fig. 11.18   Stereoscopic representation – perylene dimer A – new bonds formed are striped.

[*Refs. on p. 208*]

layer, but it is positioned away from the ring, and is assigned a chemical shift near 5.5 ppm.

The top monomer skeleton has four double bonds which are cross conjugated, and three cyclopropane rings. Two of the three carbons which are not part of double bonds or the three-membered rings are highly strained. Thus, only one carbon in the top layer of the molecule is not part of a π system of some type. This layer is believed to be the chromaphore responsible for the anomalous π-π^* absorption. The other monomer skeleton contains the naphthalene chromophore and two three-membered rings. This naphthalene chromophore is fully conjugated to only one of the three-membered rings. There are eleven new bonds in this model of dimer A. Five of these are part of cyclopropane rings. There are six cross-linking bonds between layers. Since there is one more double bond in dimer B, only ten new σ bonds are formed. Four of these are involved in three-membered rings and six connect the monomer layers. It should be emphasized that while the model fits the available observations, other slightly different models do almost as well. The material is undoubtedly a mixture of closely related compounds.

A similar analysis has been performed for the pyrene tetramers. The details appear in the original paper [9]. Fig. 11.20 presents the n.m.r. spectrum. Table 11.1 summarizes the properties. Fig. 11.21 is a stereo-scopic representation which satisfies the available data on tetramer A. The tetramer model consists of four skew monomer like layers, which are turned as much as 90° from being aligned with each other. The layers are also shifted from a vertical position with respect to each other. Therefore the model has a leaning appearance. The mass and electronic spectra indicate that the monomer layers are distinct. The σ bond structure of the molecule has not been altered but many π bonds have been broken and rearranged into three-membered rings and cross-links between layers.

The model contains six cyclopropane rings. One proton on a carbon isolated on a three-membered ring is shifted downfield to about 1.5 ppm by van der Waals interactions with another proton. Two three-membered rings are in each layer containing a benzene ring. These layers are the top and bottom layers of the tetramer. The two benzene rings are part of a styrene chromophore which is shifted to lower energy by the two three-membered rings in each layer.

There are two double bonds which have protons attached to them. These protons are over the edges of the benzene rings. The double bond containing two protons is in the ring of a monomer which has

[*Refs. on p. 208*]

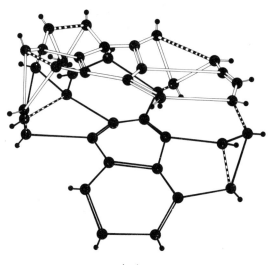

(a)

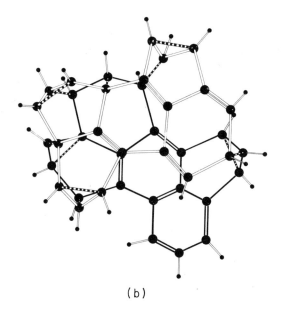

(b)

Fig. 11.19(a) Side view – perylene dimer A – new
bonds formed are striped; (b) top view – perylene
dimer A – new bonds formed are striped.

[*Refs. on p. 208*]

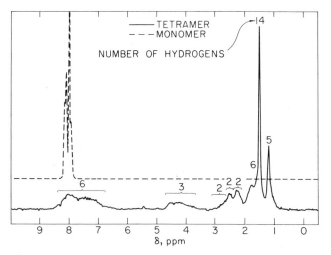

Fig. 11.20   NMR spectrum – pyrene tetramer A.

only two protons. The other olefinic proton is on a ring which has three protons. A number of paraffinic protons, shifted downfield by a neighboring double bond, are shifted upfield an equal amount by diamagnetic effects from a benzene ring or combinations of benzene rings and double bonds. These protons, those unshifted by double bonds, the proton from a three-membered ring shifted downfield by van der Waals interactions, and those on both three and five-membered rings, add up to 14. These appear at 1.5 ppm. Other protons are shifted downfield by double bonds or benzene rings and are not shifted upfield very much by diamagnetic or anisotropic effects. These appear between 1.6 and 3.0 ppm. They total 13 protons. Eighteen new bonds are formed in tetramer A. Twelve of these are also between layers. The mass spectra for the tetramers include peaks at $m/e = 802$ to 808, 602 to 606, 400 to 406 and 200 to 206 indicating that the tetramer maintains a distinct 'memory' of its monomeric origin. It was possible to build many similar models but this one best fits the facts known about the tetramer.

The interest in these compounds lies both in the formation of a new class of hydrocarbon products and in the mechanism of the reaction. The pure hydrocarbons do not react within the available pressure range. The low energy π-π^* transition in perylene lies at 2.87 eV, while that of pyrene is at 3.67 eV. These peaks shift to lower energy with pressure, but they are relatively narrow ($\sim$0.1 to 0.2 eV), and no reasonable combination of shift and broadening would permit thermal excitation at the available pressures. The charge transfer peak for the pyrene complex

[*Refs. on p. 208*]

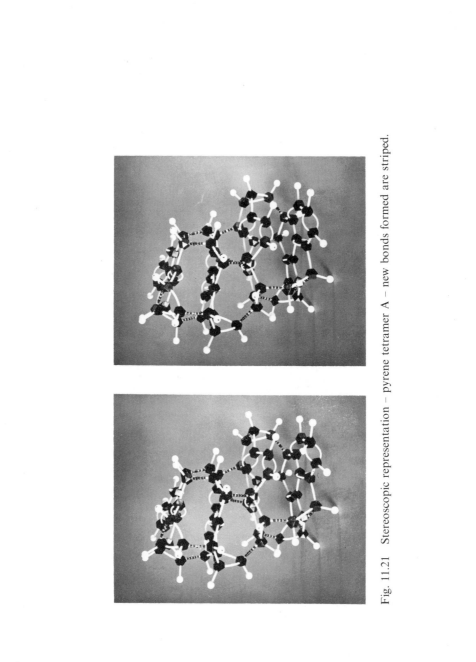

Fig. 11.21   Stereoscopic representation – pyrene tetramer A – new bonds formed are striped.

[*Refs. on p. 208*]

lies at 2.96 eV with an estimated half width of ~ 0.7 eV. The peak for the perylene complex is presumably buried under the intense iodine peak at lower energy. The presence of this iodine peak makes pressure studies of the CT peak very difficult. An application of Equation (3.22) indicates that a shift of 0.5 eV accompanied by 10 to 15% broadening or 30% broadening with negligible shift would be sufficient to give very substantial thermal occupation of the excited state. The shift studies of Chapter 6 would indicate that these conditions would easily be met in the 100 kbar range. At 1 atm, electron spin resonance measurements indicate that the concentration of unpaired spins is 5 to 7% of the hydrocarbon molecules [14]. Table 11.2 contains the results of atmospheric pressure electron spin resonance studies on a series of pellets of pyrene-$2I_2$ taken to 80 and 165 kbar for various periods. Compression to 80 kbar multiplies the unpaired spins by a factor of three. At the higher pressure the reaction reduces the number of unpaired spins. Also, at either pressure the number of unpaired spins decreases with time as the reaction proceeds. Removal of the iodine destroys practically all unpaired spins, as it does for unreacted complexes.

Table 11.2 Relative concentration of unpaired spins for pellets of pyrene-$2I_2$ complex recovered from various pressures

Material	Hours under pressure	
	0*	24
Unpressed complex	1.00	—
After 80 kbar	3.03	1.98
After 165 kbar	1.93	1.72

* Taken to pressure and released immediately.

The reactions discussed here proceed through the excited state of the complex. The reaction mechanism could go either through the charged aromatic donors or through an uncharged excited state, produced when an electron in the charged excited state of iodine returns to the aromatic molecule. Many observers have reported reversible dimer formation in pyrene and perylene systems under a variety of conditions [15]. The reversible dimers formed are generally of the Ar_2^+ type, although dimers of the configuration Ar_2^{+2} have been observed, as well as uncharged excimer formation of the type $Ar:Ar^*$ [15, 16]. Excimers involving higher excited states have also been observed [17]. There is ample evidence,

[*Refs. on p. 208*]

then, of associations between excited and unexcited aromatic molecules as well as between two excited molecules. Under pressure these excited complexes of aromatic molecules could react further to form the dimers and tetramers observed here. At pressures under 10 kbar Offen has shown that the rate of dissociation of pyrene excimers is strongly inhibited by pressure [18]. Of particular interest is the function of iodine as a catalyst. Under high pressure it provides a low-lying excited state which is thermally available at room temperature. While it may be bound to the hydrocarbon in some intermediate stage, it does not appear in the final product.

It is clear that there must be significant changes in molecular arrangement at high pressure to provide the necessary geometry for the reaction. In the present state of our knowledge concerning the geometry of the complex, particularly at high pressure, any discussion would be purely speculative.

These studies are, of course, in a relatively rudimentary stage. They do illustrate, however, the possibility for chemical reactivity as a result of the relative shift of electronic energy levels with pressure. A new class of hydrocarbons is created. It would appear in principle possible to design complexes with appropriate geometry and electronic structure to create a variety of desired products.

References

1. R. FOSTER, *Organic Charge Transfer Complexes*, Academic Press, New York (1969).
2. B. PULLMAN and A. PULLMAN, *Nature* (*London*), **199** 467 (1963).
3. G. A. SAMARA and H. G. DRICKAMER, *J. Chem. Phys.*, **37** 474 (1962).
4. R. B. AUST, W. H. BENTLEY and H. G. DRICKAMER, *J. Chem. Phys.*, **41** 1856 (1964).
5. V. C. BASTRON and H. G. DRICKAMER, *J. Solid State Chemistry*, **3** 550 (1971).
6. R. B. AUST, G. A. SAMARA and H. G. DRICKAMER, *J. Chem. Phys.*, **41** 2003 (1964).
7. W. H. BENTLEY and H. G. DRICKAMER, *J. Chem. Phs.*, **42** 1573 (1965).
8. I. L. KARLE and A. V. FRATINI, *Act. Cryst.*, **B26** 596 (1970).
9. M. I. KUHLMAN and H. G. DRICKAMER, *J. Am. Chem. Soc.* Dec. (1972).
10. T. UCHIDA and H. AKAMATU, *Bull. Chem. Soc. Japan*, **34** 1015 (1961).
11. O. HASSEL and C. RØMMING, *Quart. Rev.*, **16** 1 (1962).
12. J. A. POPLE, H. J. BERNSTEIN and W. G. SCHNEIDER, *High Resolution Nuclear Magnetic Resonance*, McGraw-Hill, New York (1959).
13. J. W. EMSLEY, J. FEENEY and L. H. SUTLIFFE, *High Resolution NMR Spectroscopy*, Pergamon Press, New York (1965).

14. Y. MATSUNGA, *J. Chem. Phys.*, **30** 355 (1959).
15. J. B. BIRKS, *Photophysics of Aromatic Molecules*, Wiley-Interscience, New York (1970).
16. A. KIRA, S. ARAI and M. IMAMURA, *J. Chem. Phys.*, **54** 4890 (1971).
17. C. R. GOLDSCHMIDT and M. OTTOLENGH, *Chem. Phys. Lett.*, **4** 570 (1970).
18. P. C. JOHNSON and H. W. OFFEN, *J. Chem. Phys.*, **56** 1938 (1972).

Units of Pressure and Energy

Pressure

$1 \text{ kbar} = 10^9 \text{ dyn cm}^{-2}$
$= 986.9 \text{ atmospheres}$

Energy

$1 \text{ eV atom}^{-1} = 23.053 \text{ kcal. g atom}^{-1}$
$= 8067.5 \text{ cm}^{-1}$

Author and Subject Indexes

Author Index

Subject Index

The most general effect of pressure is to increase overlap of electronic orbitals. An important consequence of this increased overlap is the shift in energy of one type of orbital with respect to another. In a very wide variety of cases this shift is sufficient to provide a new groundstate for the system which may have very different physical and chemical properties. Although this book discusses the physical consequences of these electronic transitions, it lays greater emphasis on the *chemical* effects and in so doing seeks to provide a common thread in a wide variety of apparently disparate observations

The first part of the book contains an account of the theoretical background necessary for understanding the observations, while the second part is concerned with the observations themselves.

This book will be invaluable to many workers in chemical physics and solid state physics. It will interest experimentalists in new topics which are barely scratched or completely untouched. Theoreticians will find here a most broad and fertile field for further analysis.

The Authors

Professor Harry Drickamer's outstanding contributions to the subject have been recognized by his winning of awards from, among others: the American Chemical Society (Ipatieff Prize, 1956), the American Institute of Chemical Engineers (Alpha Chi Sigma Award, 1967), and the American Physical Society (Oliver E. Buckley Solid State Physics Prize, 1967). He has been Professor of Physical Chemistry and Chemical Engineering at the University of Illinois since 1958.

Dr. Curt Frank completed his Ph.D. at Illinois in 1971. He continued working at the University until taking up his present position of Research Scientist at Sandia Laboratories, Albuquerque, New Mexico.